KILLING WITHOUT HEART

Related Titles from Potomac Books

Predators: The CIA's Drone War on al Qaeda
Brian Glyn Williams

Getting Away with Torture: Secret Government, War Crimes, and the Rule of Law
Christopher H. Pyle

Imperial Designs: War, Humiliation & the Making of History
Deepak Tripathi

KILLING WITHOUT HEART

Limits on Robotic Warfare in an Age of Persistent Conflict

M. Shane Riza

Foreword by Martin L. Cook

Potomac Books
Washington, D.C.

Potomac Books is an imprint of the University of Nebraska Press

The views expressed in this work are those of the author and do not reflect the official policy or position of the National Defense University, the Department of Defense, or the U.S. government.

Library of Congress Cataloging-in-Publication Data
Riza, M. Shane.
Killing without heart : limits on robotic warfare in an age of persistent conflict / M. Shane Riza ; Foreword by Martin Cook. — First edition.
pages cm
Includes bibliographical references and index.
ISBN 978-1-61234-613-7 (hbk. : alk. paper)
ISBN 978-1-61234-614-4 (electronic)
1. Robotics—Military applications. 2. Robotics—Moral and ethical aspects. 3. Military robots—Moral and ethical aspects. 4. Drone aircraft—Moral and ethical aspects. 5. Impunity. 6. Military ethics. 7. War (Philosophy) 8. Military art and science—Forecasting. I. Title.
UG479.R59 2013
172'.42–dc23

2013000748

Printed in the United States of America on acid-free paper that meets the American National Standards Institute Z39-48 Standard.

Potomac Books
22841 Quicksilver Drive
Dulles, Virginia 20166

First Edition

10 9 8 7 6 5 4 3 2 1

To my fellow combatants and warriors.
May our nation use us wisely—and may we always fight well.

CONTENTS

Foreword by Martin L. Cook ix
Preface xiii
Acknowledgments xvii

1: A Fighter Pilot's Entry into the Robotic Age 1
2: Now and When: Current and Future State of Robotics in Warfare 7
3: The Law Is Not Enough 25
4: The Spectra of Impunity in Warfare 37
5: Impunity and the Politics of War 61
6: Impunity and the Warrior 83
7: Impunity and the Future of War 109
8: AI, the Search for Relevance, and Robotic *Jus in Bello* 125
9: Radical Responsibilities 149
10: Inevitability, Persistence . . . and Heart 167

Notes 179
Selected Bibliography 203
Index 211
About the Author 221

FOREWORD

The technology of war is always changing. Combatants are engaged in a perennial struggle to achieve superiority over their adversaries and to maximize their own survival. Warfare began with adversaries in close proximity to each other struggling to control their fear and prevail. It has evolved from standoff weapons, such as the slingshot and the bow, through long-distance artillery, into modern remotely piloted vehicles ("drones"), and potentially into a future of semiautonomous or even fully autonomous weapon systems.

History provides little precedent to suggest that technological evolution in warfare can be stopped, no matter how excellent the arguments that perhaps it should. At one point the Christian Church attempted to prohibit the crossbow as a weapon excessive in its long-distance lethality. Yet crossbows became standard weapons of war. Rapid-fire machine guns turned assaults over open ground into suicide missions in World War I—and yet they just continued to evolve. Two exceptions stand out from the iron law of technological necessity (at least so far)—weapon systems that were developed, used, and then stepped back from: chemical and nuclear weapons. Both remain in the arsenals of some militaries, but the restraints on their use have held.

M. Shane Riza has written a profound reflection on one important area of rapid technological evolution in recent years: the developments in the technologies of unmanned (and perhaps increasingly autonomous) weapon systems.

While much has already been written about these developments, the range of the discussion has been fairly narrow, and the ethical frameworks limited. Traditional just war categories of discrimination/distinction and proportionality have largely set the conversation. Discrimination/distinction asks, "Can these weapons be used accurately? Do they have high probability of striking legitimate military targets?" Proportionality asks, "Can the damage done to nonmilitary objects and personnel be kept acceptably low, in light of the military value of the legitimate military object destroyed?"

Using that frame, ever-increasing reliance on such unmanned systems seems completely acceptable and perhaps even ethically preferable to almost any alternative system. They provide maximal force protection for our own forces, since the weapons operate in environments where our own operators are not even physically present. Surely if there is any ethical imperative for military and political leaders, it is that they should not put their own people at risk of death or injury if it can be avoided. The long-loiter capability of unmanned systems gives operators the ability to observe potential targets for hours before launching a weapon. This, the argument goes, dramatically improves the chances that, when they do launch, it will be toward a legitimate target. Further, the precision of the guidance systems allows weapons of much lower explosive yield. This should greatly reduce collateral damage and confine kinetic effects of the weapons to as small a blast radius as possible. Given these many considerations, what's not to love about unmanned systems? Perhaps war can be made nearly "immaculate," and how could that be bad?

To these apparently overwhelming considerations, Colonel Riza brings the unique perspective of a deeply reflective, experienced combat aviator. The questions he raises go far beyond the fairly narrowly defined tactical just war framework. If such weapon systems allow us to destroy personnel and objects with nearly perfect impunity, is it even war? Do "unusually useable" weapons (i.e., weapons that can be used with near-zero risk to our side) inevitably lower or even erase the high bar that just war theory sets in the first place for resorting to military force? Why should war be a last resort, when all our side puts at risk is some hardware? Does that turn the entire debate about entering war from a grave and solemn moral conversation to one of cost-benefit analysis? Have we begun to think through the law of armed conflict implications of these weapon systems? He notes that an adversary who had the capability to target a remotely piloted vehicle operator in Nevada would fall squarely within the legal permissions of war. He invites and urges us to think through the precedent-setting implications of present-day American use of these weapons. For example, if we establish the practice that the United States, on the sole authority of our executive branch, can reach into the borders of sovereign states to kill individuals (including our own citizens) by drone strikes, what happens when that capability is available to many other states and even to non-state actors? Have we really taken into account the long-term consequence for a peaceful future with present adversaries if our means of attacking them are perceived as deeply dishonorable and cowardly? Isn't it important to respect the humanity of our adversaries, even as we fight them, if we have any hope of returning to be "better peace" at the end? And most profoundly of all, he questions what these developments do to the entire moral meaning of being a warrior whose actions have meaning and nobility precisely because they manifest a kind of human prowess and excellence in the face of fear and danger.

Colonel Riza's is an important and eloquent plea that we find the time and space to think more deeply about these larger issues before the juggernaut of technology renders the entire discussion irrelevant. Further, should that "look before we leap" plea fall on deaf ears, this important work raises the kinds of questions we might need to revisit even after we let the genie out of the bottle. History might suggest we are likely to rush blindly down the path of perceived technological imperatives. But there may come a future moment when, as happened with chemical and nuclear weapons, we realize only after the fact that the military world we've made is so frightening and horrific that we are finally prepared to reconsider the use of technologies we've already developed.

In any case, Colonel Riza has made a profound and unique contribution to a discussion that, perhaps, has been too narrowly framed. Philosophers, theologians, and politicians are well advised to listen carefully to the voice of the warrior philosopher.

Martin L. Cook
Adm. James B. Stockdale Chair of Professional Military Ethics
U.S. Naval War College

PREFACE

The idea for this book grew out of a meeting held a few years ago on an air base in the rural north of Japan. It was between four squadron commanders, our group commander, and an Air Force colonel from an office on the Air Staff. He was there to ask us to consider sending our top performers to unmanned aircraft, what the public has come to know as "drones," because of the very important missions these weapon systems were doing. His pitch assured us that our top picks would go to certain squadrons of our choosing, but he was unable to offer the guarantee in writing. He was also unable to say whether these volunteers, if there were any, would be able to return to the fighter many had dreamed of flying since they were children. I asked one question. "What legal ground are my guys going to be on?" In other words, how would those I sent to do this work be judged under the laws of war? How would they be viewed through the lens of the just war tradition and the filter of history? His answer did not instill confidence. He did not say, "They are fully protected under the law of war." What he said was, paraphrased, "We think we're on pretty solid ground."

This one conversation is a microcosm of the issues surrounding armed unmanned vehicles. They dredge up deep service culture conundrums. We are discovering the moral, ethical, and legal nature of these kinds of systems without the due consideration required. The story highlights interesting challenges in organizational structure and training for war fighters, but most of all it highlights the uncertainty of what this brave new world is really all about. That meeting and other discussions with my fellow commanders at the bar, on patios, and in each others' kitchens started my quest to understand the coming robotics revolution (if it is one), why it might be different than other revolutions in warfare, what it may mean for the future of war, and what it may mean for the future of the warrior.

Technology has aided warfare since the very first human to fashion a spear launched it into the air by the power of his arm. As we have seen, these kinds of technologies, even in their simplest forms, increase the distances at which combat could be conducted. They increase the physical distance from the killing, and in instances where one side wields a greater capability than the other, they reduce the risk for the side with the technological advantage. All of these are accepted and acceptable in warfare. There is no desire or requirement for a fair fight, but there is something going on that may be changing the entire calculus. This change could shake the moral foundation of warfare.

Today three technological trends are shaping the way war is fought. These trends are driving a desire and capability for near total impunity. Impunity shakes the moral foundation of war, inverting the traditional risk paradigm by transferring risk to noncombatants. It affects the politics of war, the warrior, and warfare's future through its impact on the "conversation" among combatants engaged in the game of mortal combat. The advent of armed unmanned vehicles and the possibility of lethal autonomous robots are speeding these changes and threatening to impose on us a new template for war before we have even tried to understand it.

I have attempted to explain these trends and draw conclusions about the moral, ethical, and practical limits of robotic warfare. In doing so, I have surely left much out. Where it was relevant to do so, I have attempted to point out these shortcomings in the text. Where they are not highlighted, I either neglected them due to perceived scope, or I simply missed them. I have been aided by many people during the course of this project, people who have attempted to make a researcher, ethicist, lawyer, and writer out of an old fighter pilot and warrior. I'm grateful for their attempts, but I take full blame for any errors in my understanding that may be conveyed within.

When I began this work, my goal was to further the discussion about the deeper issues surrounding armed unmanned and robotic warfare. As these kinds of works tend to do, it took some unexpected turns. At its core, this book is a defense of the warrior ethos, and it questions our ability to maintain what little of it may be left in the coming robotic age. It questions whether it even is or will remain a necessity. I am well aware this may put me in the same category as Army officers who could not let go of the horse in favor of the armored tank. As philosopher Albert Camus said in disdain for those discounting the contributions of others in different theaters during World War II, "When everything has been made vile and sordid, they still try to establish an order of merit. That is how they survive."[1] Perhaps this will be viewed in the same light—a fighter pilot just trying to survive. It is a plausible criticism. To be sure, my motive—both within these pages and in whatever time remains for me in the profession of arms—is in the preservation of a select group that knows and understands the awful complexity of war and the awesome responsibility of those who decide for and lead in it. If this work better allows us to talk about these things as we attempt to understand our changing way of war, then it will have been worth it. I can then someday safely retire to a house on the river and spend the rest of my life, in spite of Camus, remembering how great I once was. I can fondly remember the days when killing, and debates about its justice, took heart.

ACKNOWLEDGMENTS

First, I have to thank my ever-patient family, who rightly thought our year in the Washington, D.C., area would be one of traveling, seeing the historical sites that played such a role in creating this great nation, and spending plenty of time with their husband and father. They did not know I would decide to write a book during my year of "academic recuperation" at the Industrial College of the Armed Forces—now the Dwight D. Eisenhower School for National Security and Resource Strategy—on the beautiful campus of the National Defense University at Fort McNair in Washington. We might not have seen all the sights of D.C., but having never been assigned to the Pentagon, I'm assured I will likely get another opportunity to live and work there.

It would be impossible to express enough gratitude to my adviser, retired fighter pilot Stephen Randolph, then assistant dean of faculty at the college and now the State Department historian, for his guidance when appropriate but most of all for his trust in my ability to see this through to fruition. He gave me the space to explore this complex subject and was as enthusiastic as I was at what I was coming to understand.

I also have to thank Kenneth Moss, then chair of the National Security Studies Department and professor for my class of the same name, for his early interest and encouragement in this project. I am also grateful for his careful reading of drafts of this work, often delivered to him haphazardly, and for his insight as an expert in the politics of war. I learned much in his class, but I learned more talking to him about the issues discussed in this book. Time spent in the bowels of the far southeastern corner of the first deck of Eisenhower Hall was never wasted.

Thank you too to all those I have served with and learned from. A special thanks to Col. Stephen Platt, my friend of more than two decades, without whom portions of this work would have been far less human. He understands far better than I the personal nature of killing in war. We could not have known it at the time we were skipping out on studying for some engineering final exam to go rock climbing, but we are forever linked by an event that completely altered the way I look at my chosen profession and its ultimate cost. Thanks also to Col. Jay Aanrud and Col. Joe McFall who, whether they know it or not, were there at the inception of this idea and whose deep discussions on the issues expounded on here—solving the world's problems on the back of bar napkins—continue to shape my thinking. I'm proud to be in your company. Brothers in arms, all of you.

I extend thanks to those who assisted me through interviews and by simply answering my many questions. They made this work immeasurably better than I ever could have done on my own. They include Chris Lafferty; Col. "Max" Maxwell, USA; Ronald Arkin; Mary Ellen O'Connell; Missy Cummings; George Lucas; and the great many I cannot name due to promised nonattribution. Some of you know who you are. For the others . . . well, that is partially the point, I suppose.

Finally I would like to thank Hilary Claggett, Melissa Jones, Laura Briggs, Sam Dorrance, Don McKeon, and all those at Potomac Books for seeing the possibilities of this work and for believing in its contribution to the much-needed discussion on the consequences of war by robotic or autonomous means. I can only hope this book adds to our continuing conversation about what is right—what is just—in the violent cauldron of the seemingly necessary human activity of warfare.

1

A Fighter Pilot's Entry into the Robotic Age

Act as if the maxim of thy action were to become by thy will a universal law of nature.

—Immanuel Kant[1]

We might say that war is a philosophical creation. Long before philosophers are satisfied with it, however, soldiers are bound by its canons.

—Michael Walzer[2]

There are human and inhuman warriors, just and unjust wars, forms of killing that are necessary and forms that dishonor us all.

—Michael Ignatieff[3]

In the early fall of 2008, I found myself walking around an air base that, over the course of several years, I had spent many deployments attempting to keep out of commission. In the months following Operation Desert Storm in 1991, Saddam Hussein used his air force to put down rebellious Kurds in the north of Iraq and Shia in the south. The response of the United Nations was to invoke no-fly zones over much of Iraq in the hope of keeping Saddam's still formidable military contained, reducing the internal threat to Iraqi citizens and the external threat to the rest of the region. If the overall UN strategy of Operation Southern Watch and Operation Northern Watch was deterrence—or at least dissuasion—the object of every red-blooded fighter pilot was getting a MiG kill. What better way to deter a dictator from using his air force than to "redistribute" the parts of several of his French- or Soviet-made fighters over large portions of the Iraqi countryside?

Unfortunately for those young pilots engaged in what turned out to be pure diplomacy, Saddam lost the stomach for directly challenging coalition—mostly U.S.—air superiority early on. When Operation Northern Watch was still in its own youth, an F-16 enforcing the no-fly zone shot down an Iraqi MiG-25 that had wandered too far north in a vain attempt to reclaim Saddam's sense of sovereignty.[4] For the remainder of those operations in the eleven years leading up to the launch of Operation Iraqi Freedom in 2003, U.S. Air Force (USAF) fighter pilots hoped in vain for MiGs or Mirages to launch from such places as Balad, Al Taqqadum, Kirkuk, and Irbil so they could demonstrate their prowess and the overwhelming superiority of U.S. technology in air-to-air combat.[5]

In those days, flying "in the container"—the airspace above Iraq—was a risky proposition. Coalition pilots were fired upon regularly, and Saddam offered a bounty for any U.S. aircraft shot down while enforcing the no-fly zones.[6] For these reasons, and because the U.S. military had not occupied a country on this scale since the end of World War II until Operation Enduring Freedom and Operation Iraqi Freedom, landing at Balad Air Base with the intention of staying several months was a surreal experience. That initial feeling of being somehow misplaced was relived again and again as I went to staff meetings in the offices of the Baathist former base commander and maintained my own operations out of a former MiG-23 squadron's headquarters. I stood on top of an unused ammo bunker gazing out at the sunset over the vast western desert of Iraq and considered how my counterpart, that MiG-23 squadron commander, had gone about his daily business unable to fly over large portions of his own country. I wondered how he dealt with the fact I dealt with on a nearly daily basis—how to protect your people from those living outside the base who disliked your presence so much they thought it fine to lob a few mortar rounds your direction every now and then.[7]

It was during one of these musings that a sound I had never heard before made me jump and turn to face what I thought just might be my end. The noise from a Counter Rocket Artillery Mortar (C-RAM) Gatling gun is what I imagine the burp of a mighty dragon might sound like. It is a low, guttural sound with a slow and very limited rise in pitch as the 20mm barrels spin up to full speed. Only when the initial shock to your hearing begins to subside (in the temporally distorted fashion that threats to life and limb take), do you know the C-RAM's true sensation. You *feel* it reverberating through your chest in a way that defies soldiers' and Airmen's attempts to drown out the world around them by sticking buds to MP3 players deep into their ear canals. When the C-RAM fires, no one misses it.

The C-RAM is a modified Phalanx close-in ship-defense system originally designed to knock down antiship missiles. It fires 4,500 rounds per minute, or about seventy-five per second. What makes it a C-RAM and not a Phalanx is modified software allowing it to engage rockets, artillery, and mortars—as its name states—and its deployment on land as garrison defense. Its 20mm rounds are designed to self-destruct in order to reduce collateral damage, particularly when used in urban environments.[8] When seen at night or in low-light conditions such as an Iraqi sunset, this feature gives it the look of a dragon to match its sound as hot rounds spew forth from the mouth of the gun then explode perpendicularly to the line of fire as if encountering and attempting to envelop some unforeseen structure in the distance.

The Phalanx and the C-RAM are capable of automatically detecting incoming rounds, determining if the trajectory is a threat, and firing to counter the threat.[9] By many definitions this makes the C-RAM a robot—a programmable machine, with at least some minimal autonomy, that can sense and manipulate its environment.[10] For a fighter pilot steeped in the aerial tradition of carrying the fight far afield to interdict aggressor forces and in conducting swirling battle in defense of some strategic prize, such is the surrealism of what is twenty-first-century warfare. I was standing on ground that at one time would have meant that mission failure had turned me into the propaganda pawn of a dictator. Now I was subjected to the random threat of indirect fire while being protected by a robot far beyond my control and possibly acting of its own accord. In the half-light of dusk near the banks of the Tigris River and in the cradle of civilization, it seemed like an embarkation to a whole different world.

In reality, I had encountered robotics in warfare much earlier. As a wide-eyed second lieutenant enamored of the idea of becoming a fighter pilot, I had watched Operational Test and Evaluation pilots and missile engineers coming to grips with the problems of trying to make an infrared (IR) air-to-ground missile behave the way the software engineers said it should have. The missile had an ability to track a particular part of an IR scene, enable the guidance logic to steer the missile toward that point in the scene, presumably fly directly into that point, and, due to the force of impact, trigger the explosive train that would destroy the target. In some sense (and something I never would have considered at the time), this missile was a robot: it had the ability to sense, act, and alter—quite violently as it turned out—its environment.[11]

Hundreds of things have to go right for all that to occur, and thousands of lines of computer code have to be written just right to make all those hundreds of things go right. A weapon built by one contractor must effectively communicate with an aircraft built by another in order to display the proper things to the pilot and enable targeting. The aircraft must send the correct signals to the missile to allow the pilot to control the weapon and accurately signal its release. The rocket motor must fire and burn correctly. The sensor must be able to continuously track an ever-changing scene as the missile closes on the target, and all points around where the weapon is aiming expand outward from that point. The guidance-and-control system must allow the sensor to remain on the target through the violence of sudden acceleration and the uncertainty of taking flight on its own away from the stability of the launch platform. The launch transient, as it is known, is particularly nasty, but it wasn't the problem the test pilots and engineers were facing.

The string of events described above is what we technological warriors facetiously call the "consecutive miracles," and it is a part of every technologically advanced weapon system in operation in the world today. In effect consecutive miracles are now the norm. The problem in this missile's case was that somewhere along that line of unbroken and mutually reliant miracles, one of them was no longer happening. The chain was broken somewhere in those thousands of lines of code, leading an engineer to remark, "At least it's only software" (as opposed to a hardware or design flaw that often costs more in time and money to fix). The phrase became so common and so annoying to my boss, the man tasked with making all the consecutive miracles on all the fighter planes in the USAF inventory work all the time, that he once grunted, "The next guy that tells me 'it's just software' will find himself hanging by his. . . ." This particular problem turned out to be "unsolvable," which meant there was not enough money and/or will to fix the software or that the system was actually performing to the standard set in the contract. The latter was the case here, since the contract never stated the missile had to hit the target; it only had to come within "*x*" feet of the aiming point. The pilots and engineers were left with testing how to mitigate the problem and still allow effective weapon employment, a relatively common solution that has so far not hampered combatant commanders in the development or execution of war plans in any significant ways.

In my time in that test unit and before I became a fighter pilot in my own right, I observed numerous tests of various systems sensing, acting, and altering environments. As with any technological pursuit, not all worked perfectly. Some worked well enough to be released to the field. Some required significant rework. Many had software problems, but they were never "just software" as long as the boss was within earshot. Some never made it to the field at all. This process—conceive, build, test, deploy—is as old as warfare itself. Someone had to dream up the longbow. Someone was the first to climb aboard a horse draped in armor and attempt to ride at full speed while wielding a sword. I was lucky enough to see the process very early in my career and before I was a war fighter, when I naively still thought only in chivalrous terms.

What follows is not an indictment of technology in warfare. Others have plowed that ground, and it would be the height of hypocrisy for me to indulge in it. In any case, I am no Luddite. We are all beneficiaries of our technological form of warfare—it has made the U.S. military the strongest the world has ever seen. The USAF recognizes its reliance on technology and embraces its technological nature. It was born from technology, and it is technology that ensures its future. Airmen do not shy away from what author and air power commentator Carl Builder calls this "circle of faith."[12] However, it is possible we are approaching a red line, a line beyond which the world becomes a far different place than it is now as we safely stand on this side of the line. Advances in information-processing power are making greater degrees of robotic autonomy possible. As author Armin Krishnan points out in *Killer Robots*, machines in the form of land and sea mines have been able to kill automatically for a long time, but we are nearing a point when autonomous robots can *decide* to kill—and that is something altogether different.[13]

The question, then, is not about how we arrived at our technological way of war or even if it, in and of itself, is appropriate. The question is—and it has likely been lingering since the days of the longbow, certainly for at least the past sixty years—is there a point when our technology robs us of what it means to go to war? Is there a point where our distance, both literal and figurative, and our ability to engage in battle with more and more impunity alter our concept of what is right and wrong in war? Should it? What are the moral and ethical impacts of emerging technologies that allow nearly full impunity on one side while its opponents on the other still suffer the terrible human costs of ever more destructive modern warfare? And perhaps most important, to what degree is the U.S. defense establishment considering these most fundamental questions about the very nature of war and how the pursuit of technologies farther removing humans from the field of battle impacts our views of it?

I am no longer that young fighter pilot anxious to test my skills in the crucible of combat. I am no longer in denial about the role technology has played in the development of those skills and the manner it aided my every move. I am comfortable in the knowledge that my mastery, such that it was, of the technology at my fingertips successfully took me into battle and brought me back. What I am concerned with these days, much to the surprise of a once twentysomething lieutenant engrossed in the green glow of an IR missile's seeker video, is far more foundational. I am concerned with the consequences of decisions leading to the fielding of increasingly autonomous weapon systems with ever more distance and impunity for the operator and with greater destructive force for every dollar spent. I am concerned the last dying embers of a warrior culture are about to be snuffed out in favor of a system of warfare that just might destroy war's very meaning and context as a human endeavor. And I'm concerned we are doing it all at breakneck speed, without giving it the thought it so rightly deserves.

If author Peter Singer is right, as he claims in *Wired for War*, and we are in a robotics revolution, it might not surprise very many who have been keeping up with the headlines from our long and far-flung wars or tuning into the futuristic recruiting advertisements for the U.S. military. What is striking however, as I look back on over twenty years of uniformed service, is there was never a time in my career, even if I did not recognize it, when robotic technology was *not* involved in my experience of the profession of arms. If this revolution snuck up on someone whose job has had him literally immersed to the elbows in the technology of modern fighter aircraft, it just might have done the same to others whose job it is to make policy about how to acquire and use that technology—and those whose job it is to advise them. That ought to be just slightly scarier than the sound of a C-RAM splitting the calm of an otherwise beautiful Iraqi sunset.

2

Now and When: Current and Future State of Robotics in Warfare

It shall be the goal of the Armed Forces to achieve the fielding of unmanned, remotely-controlled technology such that — (1) by 2010, one-third of the aircraft in the operational deep strike force aircraft fleet are unmanned; and (2) by 2015, one-third of the operational ground combat vehicles are unmanned.

—Public Law 106-389, National Defense Authorization Act for Fiscal Year 2001[1]

You know the only time you need a roadmap? When you're lost.

—U.S. Navy staff officer, UAS Panel, Robotics and Autonomous Industry Study Seminar, Industrial College of the Armed Forces, February 2011

Sen. John Warner of Virginia introduced the language above into the National Defense Authorization Act (NDAA) for the fiscal year beginning in October 2000. He did so for two reasons. First was his concern over the growing intolerance of casualties in war and what it might mean for U.S. foreign policy. Second, in what seems paradoxical and will be addressed later in discussions on the effects of robotic technology and what it means to be a warrior, he wanted to create incentive for young Americans born in the information age to enlist in a military on the cutting edge of technology as a means of learning skills that could be put to use later in the civilian sector.[2] As Peter Singer describes it, the act created the "demand side" for military robotic technology, forcing a somewhat reluctant, or at least ambivalent, military to pursue unmanned technologies in earnest.[3] Coming ten years after Desert Storm and the so-called Revolution in Military Affairs, the idea that precision weapons had so altered warfare that it was a completely changed concept, and directly on the heels of Operation Allied Force in the Balkans (where ground troops were ruled out and civilian and military leaders seemed unwilling to risk the lives of American military personnel), the context is important.[4]

The sum of military experience in unmanned systems up to the point when Senator Warner gave the armed forces a congressional mandate to dive headlong into an unmanned future is characterized by limited unmanned aircraft system (UAS) use in Vietnam and Desert Storm[5] and small contracts for explosive-ordnance disposal robots granted by the Defense Advanced Research Projects Agency (DARPA).[6] They were seen as nicely progressing along the spectrum of risk reduction in technological warfare that has been its vector since the beginning. In this sense, it seems logical to assert that unmanned systems were viewed as evolutionary systems. Sure, they would alter some of the means in warfare. They might alter the cultures of the services to some degree as well, but they were just the next logical step for a military that had proven wars could be won, and could be judged as justified, without the need to risk the lives of American service men and women. The moral implications seem to have been assumed. After all, what could be more moral than saving the lives of our service members? The year 2010 has been relegated to the history books. Where are we along the path the senator laid out?

It turns out wars are far more efficient at creating demand than obscure subsections of public law—even when that law is the NDAA. Following the events of September 11, 2001, the national defense budget grew 74 percent from 2002 to 2008, not including the "supplemental" costs of funding the wars in Iraq and Afghanistan, operations that have never been included in the annual defense budget.[7] During the same period, spending on ground robots roughly doubled each year, while funding for UASs like the Air Force's Predator and Reaper aircraft "grew by around 23% each year."[8] The unmanned ground systems inventory grew from virtually zero in 2001 to over five thousand in 2006.[9] Singer notes that although not a single robot made the run from Kuwait to Baghdad in 2003, by 2010 twelve thousand unmanned ground systems roamed battlefields and training grounds, joined by over seven thousand unmanned aircraft.[10] Though the services have not achieved the mandated mix of manned and unmanned systems in the ten years following the fiscal year 2001 NDAA, they have—with the "help" of the wars in Iraq and Afghanistan—certainly moved toward its intent.

Whether the services have done so deliberately and with eyes wide open is in doubt.

The Roadmap to a Robotic Future and Its Current State

In the "have to have it now" environment of combatant commanders' urgent-need requests, many of these systems went to war without the robust testing requirements of the traditional acquisition process.[11] The proliferation of robotic systems also outpaced everything from planning for future acquisitions, to doctrine, to personnel. A senior civilian in the Department of the Air Force, when asked about the plan for all the personnel pouring into the Air Force's Predator and Reaper programs when the wars wound down, replied, "There is no plan."[12] In an effort to recapture a sense of deliberate planning for the future of unmanned systems, the Department of Defense (DOD) issued a plan for unmanned systems in 2007 called the *FY2009–2034 Unmanned Systems Integrated Roadmap*: "The Roadmap lays out a vision in terms of potential missions that could be performed by unmanned systems, the desired functionality and performance needed by the systems to perform those missions, and the technology advancements needed to achieve such performance."[13]

The DOD's *Unmanned Systems Roadmap* identifies four capability needs that unmanned systems should fill. The capability needs are prioritized as (1) reconnaissance and surveillance, (2) target identification and designation, (3) countermine and explosive ordnance disposal, and (4) chemical, biological, radiological, and nuclear reconnaissance.[14] Nowhere explicitly stated in the department's priorities for the next twenty-five years are lethal, direct-combat, autonomous systems. Its priorities are along the lines of what most unmanned systems are doing today, yet its goals are a bit more far-reaching. A deeper reading of the document shows the department's clear desire for increasing autonomy and for the systematic growth of autonomous systems. The document also clearly states the desire to maintain a human decision-making element in lethal use.[15] The *United States Air Force Unmanned Aircraft Systems Flight Plan, 2009–2047* refers to this as "man on the loop," denoting a slightly more detached supervisory role above the level of discrete tactical decisions or level of control as the term "man *in* the loop" connotes.[16] What follows is a brief characterization of current defense-related unmanned systems.

UASs are the clear face of military robotic systems today, though those flying and employing them may not agree with their characterization as robotic. At any rate, UASs support combatant commanders' needs for "persistent and highly capable intelligence, surveillance and reconnaissance"[17] and have truly come of age in the post-9/11 era. They have an around-the-block or over-the-hill capability to provide situational awareness to ground forces like never before. A recent speaker at the Industrial College of the Armed Forces (ICAF), a premiere senior service school for the military's future leaders, remarked [paraphrased], "Sure a soldier or Marine could peer around that corner, but if we have the capability to do it for him, why should we expect him to?"[18] UASs have been successful in prosecuting high-value targets both kinetically and through surveillance during operations in Afghanistan, Iraq, Yemen, and Somalia. The *Unmanned Systems Roadmap* states that "all Services currently employ a number of different systems across the spectrum from large to small UAS."[19]

Unmanned ground vehicles (UGVs) are not as prevalent and do not currently receive the level of funding of UASs. Approximately six thousand UGVs are deployed by Central Command in the Middle East, Central Asia, and South Asia. Central Command's urgent-need requests for UGVs are numerous. While some have been filled by modifying current systems, others are beyond the scope of current technology. The *Unmanned Systems Roadmap* focuses department and industry efforts to provide capabilities the combatant commanders require. Special Operations Command, Northern Command, and Pacific Command all have defined requirements for UGVs in mission areas specific to their theater needs. Additionally, DARPA Challenges, technology competitions for universities, research and private organizations, and citizens—specifically the Urban Challenge in which teams compete to complete an urban obstacle course with autonomous UGVs within a six-hour time limit—have pushed sensor capabilities for use on future UGVs and possibly unmanned systems in other domains. Industry has also taken note, and eighty firms and universities responded to a DOD request to form the Robotics Technology Consortium, which should "inform future investments into ground robotics technology development and better focus industry independent efforts to create UGVs suitable for military missions."[20]

Unmanned maritime vehicles (UMVs) do not appear to be as far along in testing and application as UASs and UGVs, but this statement requires qualification: the *Roadmap* is unclassified and is unlikely to be the full picture. Unmanned surface vehicles (USVs) have undergone testing primarily in reconnaissance roles. Unmanned undersea vehicles (UUVs) were the "workhorses of the mine clearing effort during Operation Iraqi Freedom" and were used during recovery operations during the response to Hurricane Katrina.[21] The Navy's plan is for the Littoral Combat Ship to be the first platform to fully apply the science and technology efforts currently under way and employ unmanned systems on and below the surface as well as in the air.[22]

In recognition of the goals codified in the fiscal year 2001 NDAA, the DOD sees the *Roadmap* as answering the mail of congressional requirements, stretching the envelope of performance, and identifying strengths, opportunities, challenges, and risks in procuring and fielding unmanned systems. In short, "this Roadmap is the culmination of a deliberate and methodical exercise to address the elements described above . . . [and] is the prescribed implementation plan directed in Public Law 106-389."[23] Understanding its impact on the moral, ethical, and legal nature of warfare requires delving deeper into the mysterious world of defining robotics.

What Is a Robot?

After an overnight snowstorm in 2011—a not uncommon occurrence for Detroit in February—I pulled into the parking lot at the American division of one of the world's largest industrial robot manufacturers. Our ICAF study group was about to spend the day talking business strategy, markets, and industry health with executives of a world-leading corporation at the top of its game. Approximately one hour into the presentation and discussion, a colleague asked an interesting question: "Could you define for us what you mean by 'robot'?" Silence filled the room as these executives glanced at each other. Finally one of them piped up with a technical-engineering definition of their specific kind of robot, which began with "a three-servo arm with an independent head, at least *x* degrees of freedom." Then my pen somehow lost interest in copying the rest. It was a definition that had nothing to do with any product not designed to work assembly lines, material handling, or warehousing functions. It highlighted a salient point: there is no common definition for a robot.

We have family friends who, because of a long philanthropic streak, tend to buy our sons really cool toys. A few years ago it was radio-controlled cars. Then marshmallow guns. A recent gift was a set of little "robot" bugs. Having aptly named them for a former squadron mascot and motto, we set them in motion and watched as the two-inch machines walked toward a wall, touched it with springy antennae, stopped, backed up, and then turned away to continue on their merry way. Now both bugs spend their nights in little containers at the foot of two beds, and it does not seem odd to think they wait with great anticipation for two boys to pick them up in the morning and set them to walking again. It occurs to me that except for the marshmallow guns—perhaps representative of a clearer time in the history of warfare—many of the gifts they get are far simpler civilian curiosities representing the DOD's more expensive, capable, and purposeful toys. But are these little guys really robots? It's not as easy to answer as one would think.

There is a dichotomy, even in civilian applications, in how we think about what robots are and what they do. It comes down to the distinction between automation and autonomy. Industrial robots, for the most part, do not fit a more complex definition of robotics tending toward autonomy. Industrial robots automate processes, but most of them do not possess those things robotics engineers are now commonly speaking of when they say "robot." Nowhere is this distinction more important—and more controversial—than in the lethal decision process and the question of how much of it can or should be ceded to a machine. Land mines and machine guns automated killing in war. C-RAMs, Patriot missile systems, or maybe a small tracked robot carrying a shotgun or assault rifle with the ability to select and fire on targets of its own choosing might just become autonomous killers. The definition of a robot is important; the difference between automation and autonomy is more so. The question of how and whether a robot decides to kill, well . . . that is even more important still.

In every technological breakthrough in warfare up to this point, humans have controlled lethal systems. Only humans have decided when to take other human life. As I will argue, this is essential in order to maintain the moral, ethical, and legal foundation of warfare as a *uniquely* human activity. Officially the DOD is clearly concerned with ensuring humans are involved in lethal decisions for now.[24] According to its *Unmanned Systems Roadmap*, "For a significant period into the future, the decision to pull the trigger or launch a missile from an unmanned system will not be fully automated, but it will remain under the full control of a human operator." It further states, "Many aspects of the firing sequence will be fully automated but the decision to fire will not likely be fully automated [read 'autonomous'] until legal, rules of engagement, and safety concerns have all been thoroughly examined and resolved."[25] These two sentences shed light on where the DOD sees emerging technologies going, and they serve to focus the contentious issues associated with these technologies in warfare. Automation and, more importantly, autonomy in weapon systems might be where the revolution lies. It may be where the red line of what is appropriate use of technology in warfare lies as well. As Armin Krishnan warns, there is a fundamental distinction between a machine that kills—even automatically, such as is the case with mines—and a machine that can decide to kill.[26]

Automation and autonomy are not the same things. The USAF's *Unmanned Aircraft Systems Flight Plan* highlights this by defining a future when "automation and autonomy merge."[27] The grappling at the crux of issues surrounding emerging technologies and future warfare is best witnessed in the *Flight Plan*. It states, "Future UAS able to perceive the situation and act independently with limited or little human input will greatly shorten decision time."[28] This is a generic statement that could apply to all manner of lethal or nonlethal systems, but later the document goes farther: "Authorizing a machine to make lethal combat decisions is contingent upon political and military leaders resolving legal and ethical questions. These include the appropriateness of machines having this ability, under what circumstances it should be employed, where responsibility for mistakes lies and what limitations should be placed upon the autonomy of such systems."[29] It is clear from the goals of the DOD's *Roadmap* previously listed that automation and autonomy are the frontier for emerging technologies. It also seems clear the DOD understands lethal autonomy is on shaky ethical ground. What a robot is and whether such a machine ought to have an ability to decide to take human life are key questions we must begin to grapple with now. What we have to do before we can have any meaningful discussion on the future impact of our advancing technology is come to grips with the clear distinction between automation and autonomy and navigate the all-too-unclear realm of the latter's spectrum.

The DOD's *Joint Robotics Program Master Plan FY2005* was the first and most extensive publication detailing how the military defines the characteristics of emerging robotic technology. It focused on UGVs and the Army's now-defunct Future Combat System and is the product of the Joint Robotics Program office born of congressional direction in fiscal year 1990 to rein in the services' various UGV programs.[30] According to this document, a robot is a "machine or device that works automatically or operates by remote control," and robotics is "the study and techniques involved in designing, building, and using robots."[31] This definition of a robot is very broad and would be challenged by others. For instance, Krishnan in *Killer Robots* defines a robot as "a programmable machine, with at least some minimal autonomy, that can sense and manipulate its environment."[32] Singer claims robots are "built on the 'sense-think-act' paradigm." A robot must have three "key components": sensors to "monitor the environment, 'processors' or 'artificial intelligence'" allowing decisions on how to respond to the environment, and "'effectors' . . . to create change in the world around a robot." [33] Ronald C. Arkin and Lilia Moshkina, leading researchers in autonomous lethality at the Georgia Institute of Technology, defined a robot for a project funded by the U.S. Army as "an automated machine or vehicle, capable of independent perception, reasoning and action."[34] A leading educational institution pioneering in degrees for robotic engineers adds "repeatability" to the definition of a robot.[35] Under its definition, a weapon on a one-way trip is not a robot. This adds another layer of ambiguity in definitions of robotic systems, but inherent in all of these is the concept of artificial intelligence, or AI.

The *Joint Robotics Program Master Plan* probably says it best in defining artificial intelligence as "the programming and ability of a robot to perform functions that are normally associated with human intelligence, such as reasoning, planning, problem solving, pattern recognition, perception, cognition, understanding, learning, speech recognition, and creative response. Artificial intelligence is an inherent requirement in all future robotics systems and will support a range of evolving requirements."[36] The pioneer of AI, Marvin Minsky, described it as the "science of making machines do things that would require intelligence if done by men."[37] Artificial intelligence is required of a robotic system if we accept the sense-think-act paradigm.

Let us now go on to consider automation and autonomy. According to the *Joint Robotics Program Master Plan*, automation is "the capability of a machine or its components to perform tasks previously done by humans. . . . Performance of tasks can be cued by humans or a point in the process."[38] This brings to mind the common view of automation in manufacturing where industrial "robots," for instance, operate on automobile assembly lines. They have replaced human workers, and the lines are therefore automated.

Whether auto manufacturing machines are really robots in accordance with the sense-think-act paradigm is an interesting point to consider. Whether it has any level of autonomy or can make decisions based on changes in its environment—indeed whether it can truly sense its environment at all or just pop-rivet, for example, at a certain location in a coordinate system representing real space—is widely variable across the industry and across platforms.[39] So is it automation that makes a robot a robot, or is it autonomy as Krishnan claims?

The *Joint Robotics Program Master Plan* defines autonomy as "a mode of control of a UGV wherein the UGV is self-sufficient. The UGV is given its global mission by the human, having been programmed to learn from and respond to its environment, and operates without further human intervention."[40] Krishnan defines autonomy as the capability of a machine for unsupervised operation.[41] Arkin and Moshkina define an autonomous robot (they do not specifically define autonomy on its own) as "a robot that does not require direct human involvement, except for high-level mission tasking; such a robot can make its own decisions consistent with its mission without requiring direct human authorization, including decisions regarding the use of lethal force."[42] These definitions, inevitably it seems, lead to other definitions and concepts.

There are two important concepts to draw from the definitions of autonomy. The first is autonomy seen as a level of control of a robotic system—the manner in which an unmanned system receives instructions that govern its actions.[43] The second concept is that the autonomy discussed here is technical in nature. It is separate from the philosophical sense of autonomy, which produces a moral agent. The concept of moral agency will be a key to understanding the limits, if there are to be any, of autonomous robots in warfare. Robot autonomy, in the technical sense, is solely concerned with its "capability for unsupervised operations."[44] This purely technical view of autonomy is important to keep in mind in further discussions about robot autonomy and levels or modes of control.

There is a growing understanding of autonomy that sees it far more as a continuum than any specific definition. The truth is no one really has a good idea. Senior corporate officials of the Association for Unmanned Vehicle Systems International (AUVSI), the most prominent trade organization for robotics and unmanned systems, recently recognized defining industry terms as a primary goal.[45] The 2011 International Society for Military Ethics conference addressed the same issue as a worthy project in order to make the language usable.[46] It is hard to have a meaningful conversation when the vocabulary is not consistent.

A definition of a robot as a machine that operates automatically or by remote control is at once too broad and too narrow. It's too broad because it does not say enough about the sense-think-act paradigm. My coffee maker works automatically, but it has no capability to sense its environment. It does not know whether there is coffee in the filter or whether there is water in the reservoir. It does not think about or perceive anything. It cannot reason that turning on its heater in the absence of water or coffee might destroy its capability to perform the only function for which it exists. It does not attempt to solve the problem or understand consequences. It has automated my coffee making—that much is true. It is very valuable to me, but my coffee maker is a simpleton as machines go. It is not a robot. The kind of technology capable of altering the fundamental nature of war has to do more than simply automate it—it has to operate with some level of autonomy and, consequently, provide a heretofore unknown degree of impunity for its human handlers. The definition above is too narrow as it neglects levels of autonomy beyond simple remote control. We need another definition of a robot and an expanded understanding of machine autonomy.

A new definition of robot should include elements of Singer's sense-think-act paradigm, backed up by others such as Arkin and Krishnan's requirement for a degree of autonomy. There ought to be some sense of repeatability too, but as long as the machine does more than one sense-think-act cycle—as is the case, for instance, in the flight profile and navigation of a modern cruise missile which is a moderately autonomous machine on a one-way trip to a violent end in alteration of some part of its environment—it should not be excluded from the realm of the robots. All of these characteristics include some form of AI, inherent in all future robotic systems. Equally important is an explicit understanding that even in remotely controlled systems (the level of control with the highest human decision-making authority and action), if there are algorithms that aid in target identification or selection that may eventually lead to lethal action (such as pattern recognition, electronic identification, face recognition software, or biometrics), these are AI inputs to a human decision-making loop that might be operating at a higher level of autonomy than the rest of the machine. These systems are on the battlefield today, and their presence already has ethical implications. They are a part of systems with a level of control that may be referred to as "supervised autonomy."

Supervised autonomy is a concept of control whereby machines are programmed for some tasks and execute them primarily without human input, but other tasks are more directly managed by humans or aid AI algorithms in order to complete tasks to higher degrees of accuracy. Missy Cummings, of the Massachusetts Institute of Technology's Humans and Automation Lab, is researching supervised autonomy with an eye toward increasing the effectiveness of the human/machine combination. She works with micro unmanned aerial vehicles (UAVs), programming things such as control loops and search patterns in order to find areas where machines are good at autonomous tasks, combine them with areas where human cognition still reigns, and create a symbiotic relationship that performs better than either would alone.[47] This may be important work for the future of unmanned systems, but it is controversial. Cummings noted in her 2011 address to the International Society of Military Ethics that the DOD's frame of reference on autonomy and automation is colored by institutional bias that could lead down inefficient paths. We may be attempting to automate the things humans do better, while holding on to things machines are better at doing, such as actually flying unmanned aircraft—a direct attack on the USAF's concept of remotely piloted aircraft, or RPAs. But these are engineering debates, much easier to solve than issues that straddle engineering and other fuzzier, more philosophical realms. For now, how about defining a few terms to level the field?

For the purposes of this book, a robot is a programmable machine incorporating any degree of artificial intelligence allowing for some degree of autonomy and an ability to sense, perceive, and act in or on its environment. Along with this definition are two new levels of control. The first is "supervised autonomy"; it will capture the USAF's man-on-the-loop concept, help define already existing nonautonomous systems with significant AI algorithms aiding human decision making, and provide room for the development of control loops relying on symbiotic human/AI decision loops to increase accuracy and efficiency. This is in line with the recently released DOD directive on autonomous weapon systems' definition of "human-supervised autonomy."[48] The second level of control should be called "learning autonomy" and defined as a machine "having been programmed to learn from and respond to its environment, and [that] operates without further human intervention."[49] "Autonomous" will be defined as a robot's capacity for unsupervised operations. Additionally this work defines a subset of robotics called "autonomous weapons." These are robotic weapons whose level of autonomy is based on their ability to trigger a destructive mechanism, select and identify targets, move under power or by the forces of physics, navigate to their targets, and have an ability to self-repair and self-replicate.[50]

Flight of the TLAM and the Robots among Us

On January 16, 1991, I was nearing the end of a long six months of flying the T-37 Tweet in the first half of pilot training. I was stationed at a USAF base in Texas, the charter of which was to graduate fighter pilots for the air forces of the North Atlantic Treaty Organization (NATO). The vast majority of the instructors were fighter pilots on loan to the training command for a tour flying "white jets" (for the paint scheme of training aircraft in those days) and teaching fledglings to earn their wings. The mood among the instructors was somber, sometimes outright hostile, because all of them were about to miss out on the first big war since Vietnam. It's hard to explain to those who have not spent their adult years training for the mechanics of war, but there is a grieving when others you know go off to fight while you stand on the sidelines. There is a sense of lost time in all one did to prepare. There is a desire to test your mettle among flying iron and the smell of gunpowder. The ultimatum on Saddam Hussein to leave Kuwait had ended the day before. It was just a waiting game now; war was inevitable. My flight instructors were about to watch their buddies go off to do the thing many of them felt they had been born to do, and they were taking it out on hapless students who couldn't quite figure out how to fly that instrument approach or do that formation turn.

The next morning I watched with the rest of the world, as the United States and the coalition it led went to war. Special operations forces punched a hole in the Iraqi air defenses, F-117s were deep inside Iraq, and precision laser-guided bombs were falling toward their targets. Reporters in the Al-Rasheed Hotel in Baghdad watched the sky light up with the now familiar scene of antiaircraft fire snaking through the night and were astonished to see Tomahawk land attack missiles (TLAMs) flying by the hotel and turning to engage their assigned targets.[51] The USAF fired hundreds of AGM-88 high-speed antiradiation missiles (HARMs), and the Royal Air Force (RAF) of the United Kingdom fired a similar missile called the ALARM, or air-launched antiradiation missile.[52] Patriot missiles sat in Kuwait and Saudi Arabia protecting coalition forces from Iraqi Scud ballistic missiles and would soon be protecting Israeli citizens as well. Iraqi mines lay in wait for an amphibious assault that would never come.[53] The robot wars had begun.

TLAMs are what Hitler dreamed of with the "buzz bomb" of the Blitz. Both are jet powered aircraft with explosives in the nose designed to wreak havoc on the enemy far beyond the front lines. The difference is TLAMs know where they are, where they are going, and how to get there. They fly to their targets via complex navigation and terrain recognition algorithms. They "see" the world beneath them, are able to correct their flight paths, and fly to their intended targets with astounding accuracy after hundreds of miles. They are robotic, autonomous weapons, and in the Gulf War they even seemed to come to life. In the words of Richard Hallion, the former USAF historian, "the pencil-like missiles bobbed and weaved their way onward, their . . . guidance systems mentally comparing the terrain below to stored images . . . in the missile's memory. . . . Entering the city, the missiles literally 'read' the city's map with a video camera . . . always mentally comparing the images against stored data."[54] Why put quotes around "read" and not around "mentally"? Writing in 1992, Hallion ascribed intelligence to a weapon of war. The TLAM did not select its target, but it faithfully, and autonomously, drove toward its destiny, sensing and perceiving its environment and updating its navigation based on those readings. Once it left the tube on the deck of a ship floating in the Persian Gulf, it was never coming back. It was on a one-way trip into history, but it used an iterative process—sense-perceive-act—to locate its prey.

Patriot missiles, the safety blanket of coalition forces during Desert Storm, are capable of autonomously classifying incoming tracks as "hostile," thereby allowing crews to fire under the correct conditions of the rules of engagement (ROE). The physics of ballistic trajectory, the flight characteristics of enemy cruise missiles, and the electronic identification of friend and/or foe are easy enough for talented software engineers to program into identification matrices. Recognizing these characteristics from radar returns is a slightly more challenging problem, but it is nonetheless possible. A system such as this uses a precise radar to sense characteristics of inbound projectiles, invokes complex pattern-recognition software to classify the projectiles, and offers a solution the human operator is likely to trust for numerous reasons. Key among these reasons are the limited time to make a decision and a strong desire to live. Unfortunately, though the physics and programming techniques are sound, anomalies occur. Sometimes these anomalies cost lives, as they did during the 2003 invasion of Iraq when Patriot missiles shot down two allied aircraft, killing both crews.[55] Patriots exemplify supervised autonomy. The decisions to fire in these instances were made by humans, but their decisions were radically influenced—perhaps to the point of abdication—by basic artificial intelligence.

It is a characteristic of electronic warfare that signals with similar parametrics (the complex classification of signal characteristics such as radio frequency, pulse width, and pulse repetition interval) appear similar to systems designed to search for and classify those signals. Modern air defense systems are typically made up of several different radars or signal emitters doing different parts of the system's overall job of stopping aerial intruders. These systems may include an acquisition radar to find the intruder, a height finder for altitude information, a target-tracking radar to feed the fire-control computer with accurate data, and an illumination radar to aid missile tracking.[56] Additionally, modern air defenses are made up of hundreds of systems involved in target acquisition, tracking, and communication. The electromagnetic spectrum where these systems operate is a complex domain.[57] It is a difficult problem to find and target the appropriate signals to enable the freedom of action the U.S. military is accustomed to.

Complicating this problem is another characteristic of electronic warfare. Missiles fired at a particular signal will track and attempt to kill the emitters of signals with similar parametrics. A HARM fired by an F-4G in Desert Storm (or an F-16CJ, EA-6B, or EF-18G today) would, in the end, attempt to hit *something* with the parametrics it was sent off the rail seeking. It may, with very rudimentary characteristics of artificial intelligence, autonomously track and kill something other than what it was designated to kill.

That is disconcerting perhaps to those of us who live on the ground, but the Brits took the concept once step farther. The RAF's ALARM is an autonomous weapon fired by strike aircraft on ingress that will operate as an antiradiation missile in the presence of its intended target signal and attempt to guide toward that emitter. However, the ALARM takes kinetic electronic attack to a new level. In the absence of its target signal, it will loiter in the area and actively search for it. It deploys a parachute and slowly rides down through the air mass, preserving persistence. If it sees a target signal, it will automatically classify it as such, fire another rocket, and guide it to, and destroy, the offending system.[58]

And what of the age-old and trusty area-denial weapon, the mine? By the DOD's current definition, a mine is a robot. It is a machine that works automatically or, in the case of some antipersonnel mines, by remote control. It is not very discriminate, but it sits waiting to sense something—pressure, the presence of metal, vibration—that tells it to function by invoking its triggering mechanism. It is perhaps more rudimentary than the systems described previously, but is it operating in accordance with the sense-perceive-act paradigm? Perhaps.

The U.S. military today is increasingly moving toward using what are termed "smart weapons." TLAMs are certainly a part of this genre, but the term is most often used to describe bombs employed from various strike aircraft. These weapons allow for a degree of precision, and presumably discrimination, never before known in the history of aerial bombardment. They are equipped with laser seekers or global positioning system (GPS) receivers enabling them to fall within several meters of where they are intended to fall. In the case of the laser-guided weapons, a crew member or ground party "illuminates" the target with a laser tuned to the code that the weapon's seeker is looking for. The seeker "sees" the laser energy, tracks it, and attempts to zero out any perceived line-of-sight motion, putting itself on a collision course with the target. Guidance and control systems take the information from the seeker and literally fly the weapon to its intended target. Much is the same with GPS-aided weapons as far as the guidance systems are concerned. The difference is these weapons are given target coordinates and, never sensing the target itself, sense their relation to it enabling similar collision-course flight profiles. In some sense these weapons act according to the sense-perceive-act paradigm. In another, they are similar to the industrial "robot" performing its functions in some coordinate system essentially divorced from the things it is acting upon. These are not autonomous weapons in any real sense—it takes human action to begin and complete the targeting solution, yet these weapons exhibit characteristics of robots. Such is the shady realm of defining these emerging technologies.

The "flight of the TLAM" is illustrative of numerous weapon systems possessed by the world's technological powers today. It and its kind meet portions of the many definitions of a robot, and many are, at least during part of their assigned mission, autonomous weapons. Whether they are actual robots is less important than their position as descriptors of a trajectory in the modern weapons of war. As one leafs through the literature and the struggle to define emerging technologies, one finds characteristics of these technologies already residing in well-known and well-worn places in the armed services.

We walk in the time of the robots. Artificial intelligence resides in our weapon systems, and it flies, drives, and sails with us regularly. Autonomous killers scream across the sky at high Mach speeds or gently float down on parachutes patiently waiting for something (and the someone inside, beside, or under) to kill. Our own lethal decisions are increasingly entrusted to technological solutions, making our grip on the humanity of war more and more tenuous. These were revelations to a twenty-year veteran and student of this ultimate game called war. If it is possible we have allowed ourselves into the place we are today without much debate and introspection, is it possible we are capable of doing the same as our technological capabilities increase? Might we turn around one day and ask ourselves when we lost the human monopoly on war?[59] If experience serves, the unfortunate and resounding answer is "yes." And its ramifications are far more unsettling than the strange feeling of emotional connection I get when I see two boys put their tiny robot bugs to bed for the night.

A Roadmap to Where?

The Joint Robotics Program, its master plan, and the DOD's *Unmanned Systems Roadmap* are all evidence of a department recognizing a need to understand the technology before us, the capabilities it may allow, and some of the challenges it will surely present. This is encouraging, but work remains to be done. Among the challenges ahead are getting a handle on exactly what we mean when we talk of robotic technologies, automation and autonomy in general, and autonomous weapons in particular. Without a common language, military and civilian decision makers, the general public they serve, and industry partners willing to make all our technological dreams come true will all be talking past each other. More dangerously, important discussions on the ethics and legality of what our dreams entail will lose out in the din of simply trying to make sense of what we are saying to each other.

The DOD's vision is that "unmanned systems will be integrated across domains and with manned systems."[60] There is a trend toward greater autonomy, learning, and adaptive systems able to collaborate with other machines and humans. Robotics will trend toward greater survivability and resiliency. Unmanned systems will be able to function in harsher environments, and they will be able to perform in conditions where humans cannot. We will proceed along the scale from remote control, in which several humans are required to *operate*—actually control—one unmanned system, to an autonomous, collaborative swarm of many vehicles being *observed* by a single operator finally incapable of controlling the individual actions of single vehicles. There will be a mix of manned and unmanned systems, but large growth in unmanned systems is expected.

While there are calls for ethical and legal reviews, there is also a building inertia and nearly insatiable appetite for unmanned systems. These systems are being fielded without the proper amount of consideration and critical thought. Noel Sharkey, a professor of artificial intelligence and robotics at the University of Sheffield in England and founding member of the International Committee for Robot Arms Control, says, "There's been absolutely no international discussion. It's all going forward without anyone talking to one another."[61] Talking seems a reasonable thing to do when confronted with the prospect of autonomous machines deciding to kill humans, but then there is this from the *Unmanned Systems Roadmap*: "When the procurement of unmanned systems threaten manned systems budgets or career paths of manned systems operators, the manned systems invariably win out due to vocal and forceful remonstrations by the threatened communities. Unmanned systems offer as yet largely unseen operational capabilities, and these pockets of resistance need to be addressed and eliminated, for the overall good of the Joint Force."[62]

Assimilation follows, it seems. The assumption that unmanned machines will fulfill as yet unseen operational needs and capabilities is, at the very least, debatable. It has to remain so in order to maintain objectivity about future acquisition priorities. Quite possibly we are overestimating this technology in the short run. It is as likely we are underestimating it for the long run,[63] but its place in the defense establishment must remain open to honest and scholarly debate. There is a congressional mandate to plow ahead in unmanned systems, but there is no requirement to do so without proper thought.

The defense establishment has lived in the time of the robots for at least the last two decades. The technologies that so impressively contributed to military success in Desert Storm and set Western militaries, with the United States as their lead, above all militaries in the world are increasingly automating warfare. Their trajectory is toward doing so with increasing autonomy. It is sufficient to say that autonomy, depending on what it controls in the machine and to what extent, is where moral peril may lie. It is where humans cede decision authority to robotic systems that the moral and ethical issues become most contentious, and this is where the focus of this book lies. This is where machines, lacking any moral agency and in an ambiguous accountability environment, may one day be allowed to make lethal decisions. How we handle this vision and what it means to the very nature of warfare are the primary moral questions for the current generation of civilian and military leaders.

3

The Law Is Not Enough

Supposing war inevitable, we have to consider what rules can be laid down for the conduct of the States engaged in it, or of other States.

—Henry Sidgwick[1]

The lawyers tell me there are no prohibitions against robots making life-or-death decisions.

—Gordon Johnson, Robotics Lead, U.S. Joint Forces Command[2]

During a recent discussion in the Pentagon's E-ring on whether robots could be programmed to act ethically, a colonel, in questioning the basis of such an exercise, blurted out, "War is an amoral act anyway!" His statement was diplomatically challenged with an "I beg to differ, but the point is," after which discussion resumed in a collegial way.[3] The colonel's sentiment might be disconcerting to just war theorists, ethicists, and other watchers of the military, but it was more an expression of exasperation with how the discussion was proceeding than any judgment on the moral standing of war or of this particular officer's opinion of it. Still, it begs the question, what is the moral standing of warfare?

The Morality of War

It seems relatively clear there must be some moral basis for warfare. If it were in fact an amoral act and generally accepted to be so, there would have been no force constricting it to the extent international law has done over the last century, and we would not react in horror over what we call "crimes against humanity" when they occur during war. International law restricts and constrains resorting to war and the acts contained within it. It has become progressively more restrictive as more laws are put on the books. With the possible exception of nuclear war, the laws of war and legitimate authorities have placed increasing limits on wars and their conduct throughout the last century. Chemical warfare is seen as abhorrent, firebombing cities with conventional ordnance is no longer part of any war plan, the growing use of precision weapons limits both the scope and the effects on civilians, and the "weapon" of choice in counterinsurgency operations often involves no weapon at all. Low-flying aircraft perform "shows of force" or "shows of presence" in so-called nonkinetic effects.[4] Why have authorities and the law so constrained the use of force? Clearly, an activity that involves killing human beings brings us all the way back to our "first principles"—those things that constitute right and wrong conduct in relation to others. War, at its heart, is a moral activity.

Evidence of war's moral nature may best be seen in the lengths we go to argue for its justification. If war were amoral, such a process would be the height of futility. Yet we do attempt to justify our decisions to go to war and our conduct once the battle is joined. "We justify our conduct; we judge the conduct of others," says just war philosopher and author Michael Walzer.[5] In fact "the theory of just war . . . is, first of all, an argument about the moral standing of warfare as a human activity."[6] In a decidedly cynical view, he then states the best evidence for war as a return to first principles and for their longevity in a relatively unchanged state is the lies soldiers and statesmen tell to justify their actions: "They lie in order to justify themselves, and so they describe for us the lineaments of justice. Wherever we find hypocrisy, we also find moral knowledge."[7] In truth, because we are human, we sometimes do fail to uphold our moral principles. Sometimes that which we attempt to justify is simply unjustifiable. But this is not evidence of a descent into amorality.

Superficial arguments are often made about war's resident evil and lack of moral standing because atrocities do occur. The counterargument is crystal clear, however. As alluded to above, our shock and horror at such events confirm the human need to propose and enforce a set of rules to govern what at first seems like only chaos. Even in the most horrific activity humankind can undertake, there is a distinction between right and wrong. There are rules to guide our just conduct. Moral philosopher Martin Cook puts it this way: "The fact that the constraints of just war are routinely overridden is no more a proof of their falsity and irrelevance than the existence of immoral behavior 'refutes' standards of morality. . . . Rather than proving the falsity of morality, [it] points instead to the source of our disappointment in such failures: our abiding knowledge of the morally right."[8]

As war is a moral activity, so it has been, to this point in history, a human activity as well. Therefore those who partake in it can—in fact, must—be judged on moral grounds. From this fertile ground grow our ethical systems and eventually the laws we use to constrain our descent to war and our actions once there. Yet there are limitations even to our own attempts to govern the actions of those who decide for and execute in combat.

Limitations of the Law

Gordon Johnson's is a common refrain among proponents of armed unmanned and robotic vehicles. Helen Greiner, CEO of CyPhy Works, stated the legality—or absence of illegality—of such vehicles recently at the AUVSI 2011 Program Review in Washington, D.C.[9] This approach reminds me of my days as an Air Force Academy cadet when what the regulation failed to prohibit was more important to my quality of life than what it permitted. The legal arguments for new weapon use are not flippant, but they do ignore some underlying issues. The most important of these is, of course, that just because no one seems to think it illegal to equip robots with guns and let them make lethal choices, it does not mean one necessarily should.

The rules for employing new weapons seem sufficiently vague, but there is a general principle. The absence of a rule means employment of a weapon is subject to the general rules of war with legality decided in accordance with those rules.[10] The Martens Clause, written into the preamble of the 1899 Hague Convention II on land war, states that "until a more complete code of laws of war is issued . . . populations and belligerents remain under the protection and empire of the principles of international law, as they result from the usages established between civilized nations, from the laws of humanity, and the requirements of the public conscience."[11] While it was originally conceived for the specific purpose of ensuring the rights of occupied populations to resist, the clause has been interpreted more generally.[12] Recent writers have wondered whether it has the force to restrict certain weapons or actions of belligerents, though it has never been invoked to do so.[13] Johnson and Greiner are technically correct, but we should rarely be satisfied with technicalities. I doubt my commander at the academy would have settled for my explanation, brilliant as it might have been, about the technical vagaries of school regulations. When the recklessness and immortality of youth fade into a too-favorably-remembered yesteryear, we are left only with our decisions and the consequences they wrought. Some are mere footnotes on the pages of a life. Others are far more important—they are the deep moral quandaries of our time.

The law is useful to justify past or desired future courses of action, but it is a poor predictor of what should be done. This is true primarily for two reasons. It is a reflection of norms built over time and therefore lags current events. It is directional but reactionary. Second, because it relies on precedent and sits atop the moral, ethical, legal pyramid, it often does not reflect accepted standards of behavior until tested in the courts, international bodies, or, in our information age, by public opinion. Quite simply, the law is not enough.

The Law as Lagging and Directional

The law of war has been laid down over centuries. Based on what are thought to be common rules about how humans relate to one another and the ethical systems of thought that can be argued by logic, laws sit atop a pyramid with morality at its base. They are distilled from our morals and ethics, but they are not themselves morality. The laws of war arose in the just war tradition and began to take on a worldwide flavor at the 1899 Hague Convention, but they do not reside in treaties alone. As author Leslie Green says, "the law of war is to be found not only in treaties, but in the customs and practices of states which gradually obtained universal recognition, and from the general principles of justice applied by jurists and practiced by military courts."[14] It has been modified as new norms of behavior gain commonality and as new ways and means of war creep into the manner by which humans resort to force to solve their differences. The law becomes more structured, possibly more restrictive, as time goes by.[15] The law is directional as it attempts to constrain the human tendency toward war, yet it can only lag human "progress" as its foundation is always in the past. An excellent example is the prohibition on antipersonnel land mines.

In the 1990s, with the worldwide distribution of antipersonnel land mines exceeding 100 million, approximately thirty people being killed per day, and an equal number being injured—and recognizing most of these casualties were civilian—a loose band of some one thousand organizations set about to ban their use. With star power lent to the cause by the late Princess Diana and others, the movement grew in appeal. In October 1996, representatives of seventy-four countries attended a conference in Ottawa, and fifty states subscribed to a declaration calling for a ban. A diplomatic conference to draft the treaty took place in September 1997, and the treaty was released for signature in December of the same year.[16] The United States, China, Russia, India, Pakistan, and others did not sign the treaty, seeing militarily effective use of certain types of mines, particularly those with self-destruct features. However, many said they would sign the treaty at a later time.[17]

The Ottawa Accord of 1997 illustrates the directional nature of the laws of war. It highlights the trend toward restricting means of warfare, but it also serves to show the reactive nature of the law of war. Antipersonnel mines of the kind that kill hundreds of innocent people every year are indiscriminate by their very nature, and indiscriminate killing had been against the law of war in written form for a hundred years before the agreement—and against the norms of behavior for a millennium before that. We should have known better than to field them. If the law had been an effective tool in those decisions, the Ottawa Accord would not have been dated 1997.

"It's legal" is not a good argument for new or revolutionary technologies in war. Biological warfare was not illegal when it was introduced. There once were no laws respecting the rights of oppressed peoples to revolt. Now there are. Land mines, in all their indiscriminate glory, still litter the landscape across the globe. The problem with the law is that it rarely prohibits anything until it has been built and fielded. It is reactive. The law was certainly challenged during the last century, and indeed it held up. It is a strong argument that it can do so again. There is one example illustrating issues in relation to supposed impunity in warfare. It is not a direct analog to the autonomous weapons debate, but it is certainly worthwhile to point out the resilience of the law.

During World War I, the Germans began to perfect submarine technology, and by the advent of World War II they had a capable force. They used it to great destruction in the North Atlantic. Although there were already three treaties and accords attempting to regulate submarine warfare, after World War II a couple of glaring difficulties became clear: distinction between combatants and noncombatants and the undue risk to submarine crews in attempting to meet the obligations required of surface vessels.[18] Additionally, jurors at the Nuremberg Trials of two German admirals, recognizing the weakness of the law in the areas mentioned above, along with the fact that all belligerent parties had conducted unrestricted submarine warfare, found them in violation but refused to convict.[19] What makes this an interesting argument for the resiliency of the law as it may bear on autonomous weapons is the similarity, in the early going, of the impunity with which German submarines were able to prosecute their attacks.

But there are two differences that are deeper still. At no time did a German submarine commander abdicate his moral agency. He chose to fire or not and in doing so decided who would live and die. His choice was a moral one. He alone was responsible, which leads to the second fundamental difference. The U-boat commanders, being responsible for their decisions, were also accountable for their actions. The admirals were judged on legal grounds and found wanting, but they were also judged on moral grounds and found to have acted as their peers and moral equals had also done. They were granted the right of all combatants in the end; that is, they were found innocent for their acts of war. Interestingly, this was not because their actions were legal. In fact, they were not. More interesting, it was not because their actions were intrinsically right either. We would have said it was unjustified—morally wrong—to kill noncombatants by disregarding the principle of distinction. Yet they were not convicted. Why?

They were innocent because the court judged them for a lack of other more appropriate courses of action given the circumstances—the undue burden of risk—and in recognition that the greater good was served by noting the violation but not punishing. They benefited from their moral equality and the capability of the members of the court to occupy the space between the law, morals, and the public conscience. It is a particularly salient point. It highlights, once again, the human and moral nature of war and its consequences.

Drone operators do not yet have to consent to equal treatment in combat—though they may soon—and autonomous weapons lack moral agency and accountability. As we scurry to hide our burning hearts' desires in the law and ever increasingly seek out military lawyers to tell us what we wish for is legal, we are, as author, scholar, and politician Michael Ignatieff says, "turning complex issues of morality into technical issues of legality."[20] It is legal expediency but moral catastrophe.

Restraint in Warfare: When Legal Is Not Right

The world is just now clawing back from the worst economic downturn since the Great Depression. It now seems clear a combination of decades-old U.S. government policy encouraging home ownership, a lack of government regulation on the shadow banking industry, and good old human greed leading to moral hazard—the difference in behavior resulting from insulation from risk—was the torrent that led us into the dark days of late 2007 and 2008. The actions of investors and mortgage companies were not strictly illegal, but few could stand and argue they were right. In writing about how corporations act with regard to the law, artificial intelligence researcher J. Storrs Hall said, "Surely, we would judge harshly a human whose only moral strictures were to obey the law."[21] Perhaps. Why then are we singularly enamored of whether our ways and means are simply legal under the laws of war?

It is hard to decide to kill someone, at least for most of us. Studies on the subject of killing in warfare find over and over the resistance is so strong that men—and I suppose now women as well—fail to fire even under threat of death. S. L. A. Marshall was the first to study this phenomenon in depth. Following World War II, he found only 15 to 20 percent of soldiers on the line had fired their weapons in any given engagement.[22] Although this study has been challenged over the years, the notion that so many veterans claimed not to have fired seems significant in its own right. Similarly, of the 27,000 muskets recovered at Gettysburg, 90 percent were found to be loaded.[23] Killing is hard to do even for trained military personnel in the heat of battle. It is a paradox of military service.

In training, when there are no real bullets spitting from a Gatling gun or missiles flying off the rails, we sometimes push the "pickle button" or squeeze the trigger without thinking through the consequences of those simple actions. We call out on the radio, "Kill Hornet over No Name, fourteen thousand, left-hand turn"[24] and move on to the next simulated bad guy getting in the way of accomplishing our mission. It is hard to remain grounded in the business of death when fighting with virtual missiles. We become supremely capable practitioners of our craft without having to consider the deeper moral questions.

Our training does not really prepare us for the thought of downing a hijacked airliner with hundreds of civilians on board or dropping a bomb on an unsuspecting conscript sitting in a radio relay station in some forsaken piece of desert hundreds of miles and a day away from an incident of antiaircraft fire in a no-fly zone. I never had the "opportunity" to deliver such retribution—there is no other word for it—as the latter, but I have a close friend who did. With all the bravado of young aviators in combat, I recall asking him when he returned from the mission whether he had taken human life. The look on his face instantly changed the way I thought about our chosen profession. Without expression he said, "I think I did."

Flying the world's best fighter was no longer just a game for him. It was, quite simply, a matter of life and death. I sat in near silence with him and watched his tape. It was a perfectly executed attack, or a "shack" in our vocabulary, meaning he hit exactly where he was aiming. In a flash of tritonal and steel, the small building ceased to exist. Whoever was inside, along with whoever drove a truck into view seconds before impact, likely suffered the same fate. They never knew what hit them. I will never forget it. Warfare became real to me that night.

In encountering an equal in the act of trying to kill you, the decision to kill is less ambiguous. It is not, however, easy and should never be allowed to be so. The first reaction to another's attempt to end your life is, or ought to be, fear. For those who survive, it is fleeting and quickly gives way to anger. For those who are disciplined and well-trained, it too is fleeting and almost suddenly—shockingly so—gives way to training and the nearly rote actions of self-defense. Then the mind turns slowly in this temporally distorted world to the act of returning fire. Based on my own experience—a sample size of exactly one—there is even then a hesitation. For me, some of it was surely the fear of not getting it right; after all, I come from a world of high standards. It is an exceedingly powerful form of restraint, but I am convinced some of my hesitation was a momentary flash of the consequences that lay under my right thumb and the red button it was tensely resting on. I fired. My missile flew off into the night searching for a ground-based radar and likely landed far out in the desert miles away from its intended target. My intelligence officer could never correlate it with any information that might have signaled a hit, but I was satisfied with the decision process nonetheless. After the adrenaline wore off and a shower removed the smell of modern fighter combat—the sweat from the fear of death has a smell all its own—I was pleased with my fleeting hesitation before I hit the pickle button.

What I expect of myself is what I believe we as a society expect of all our warriors. Doing what is legal is not quite good enough. A Defense Department official recently said to me, "We don't ask our . . . privates to read Kant and ruminate on the ethics of waging war."[25] Perhaps we should. Doing right and continually questioning our lethal choices, even if just for an instant and even while standing in the cauldron of combat, is our standard. When they choose to kill, we ought to hope our privates, sergeants, and officers do so with what my friend, in rectifying the angst over what he was called on to do to that relay station, calls "reluctant professionalism." We should have it no other way.

For a democracy with a standing army composed entirely of volunteers, there is a higher standard than the law. Our soldiers must conform to the law of war to be sure, but that is just the minimum they are expected to do. There is a collective conscience impinging on the actions of soldiers, and they are judged harshly when their actions, though legal, fail to conform to that public conscience. Then there are their own consciences, and these seem to be a far more powerful restraining force.

In his book *Just and Unjust Wars*, Walzer discusses five examples of men at war who decided to forgo killing in particular instances when, for one reason or another, they just did not feel like it was the right thing to do. One example is telling. Emilio Lusso was a lieutenant in the Italian Army fighting the Austrians during World War I. During the night he crept up on the Austrian line and in the morning saw scenes that shocked his battle-hardened senses. He watched Austrians going about the kind of life he and his men had in their own trenches. As he took aim at a young officer—his peer—the man lit a cigarette, and Lusso lost the stomach to kill him. The Austrian was so seemingly oblivious to his impending death that the young Italian could not pull the trigger. Nor could the sergeant he had brought with him. Even though he admitted to himself, "I knew it was my duty to fire,"[26] Lusso had let his humanity interfere. He wrote it this way: "To lead a hundred, even a thousand, men against another hundred, or thousand, was one thing; but to detach one man from the rest and say to him, as it were: 'Don't move, I'm going to shoot you. I'm going to kill you'—that was different. . . . To fight is one thing, but to kill a man is another."[27] You have to be the most hardened realist not to be touched by the power of Emilio Lusso's story and his actions on that day.

Walzer goes to pains to point out the decisions of these men were not choices based strictly on moral judgment. All of them described the reasons they did not fire in terms like "you don't feel like it" versus "you should not," and this distinction is indeed important.[28] It was not that it was wrong to do so, but at the same time it did not feel like the right thing to do either. As Lusso rightly states, it was the duty of each of them to pull the trigger and end the lives of those they had fixed in their sights. There is no law of war that would have prevented them from doing so or convicted them of any crime had they followed through on what most understood to be their duty. Killing "naked soldiers," or ones who are simply oblivious to war at the moment gunsights fall upon them, is both legal and, in these cases, militarily necessary in that it was required for mission accomplishment.[29] Yet these men decided against it. They were able, just as the Nuremberg jurors, to occupy that ambiguous space between the law, morals, and basic humanity. They reflect the best in us, and it is the ability of that to shine through even while engaged in the worst we have to offer that is so compelling. Decisions not to kill are "rooted in moral recognition, [but] they are nevertheless more passionate than principled decisions. They are acts of kindness. . . . Not that they involve doing more than is morally required; they involve doing less than is permitted."[30] It was the last sentence that convinced me to purchase Walzer's book and display it on my desk for as long as I have the opportunity and privilege to command men and women of war. It is the second to last sentence that sets us apart, so far as we know, from most other animals on the planet, and it certainly sets us apart from any machine most of us can currently imagine.

Our Habits of the Mind

The law is a bastion for those wondering about the possibility of lethal robotic systems and justifying unmanned armed systems. It is always the second defense in any argument on the subject—the first being the supposed moral imperative for removing the soldier from risk. Armin Krishnan says, "It can be said that international law is simply unprepared for the particular ethical challenges that are posed by AW [autonomous weapons], especially when seen from a long term perspective."[31] The irony is the law is always unprepared even for the *legal* challenges posed by any new weapon or way of war. The law can only react. It has not been successful in outlawing a technology before the technology was fielded.

This is not to say it is not resilient or is somehow irrelevant. The law of war is directional; it has trended toward greater restriction, attempting to restrain both the resort to war and the means that can be then employed. Recent successes such as the Ottawa Accords give reason for hope that international humanitarian law can further reduce the scourge war brings to populations all over the world. However, there is very little reason to expect it to be anything but reactionary; that is its nature. It seems we have to build and use our destructive toys before we can understand the full consequences of our actions. We are, after all, human. Only then, and only slowly, do the norms of accepted behavior—our moral principles—tend to change. Only then do our laws begin to reflect our new norms. This is a long process, sometimes decades long. The law lags, and the advent of autonomous weapons seems likely to widen the gap between what is possible in war and what should be legal. And even then, legality represents only a minimum standard by which we converse with our adversaries in the chaos of combat.

When what is legal and what is morally sound meet with something that just does not feel right, that is where we find our humanity. Surprisingly, for some, we find it even in the midst of war. We ought to expect our warriors to be resistant to killing. We want it to be a hard decision to make, and we ought to demand that it remain so. In war, killing must be done. It must be done in order to bring the war to conclusion, but we cannot expect those who fight to relish it. In fact we ought to be horrified if they do. We expect "reluctant professionalism"; it is a very high standard. There are times even in war when men and women will decide not to fire. They will make passionate decisions even when there are principles to guide them. Sometimes, as I will discuss in later chapters, they will make poor decisions based in the passion of rage. But sometimes they will make them from the passion of kindness. An autonomous weapon knows no passion. At times that may be good; at others it might be tragic. I wonder whether we are prepared to trade away both rage and kindness for the deadly efficiency of a machine that has no qualms about killing naked soldiers or those blissfully, if only momentarily, unaware of the war around them. Are we prepared for killing without heart? One thing is certain. We will not find the answers in the otherwise necessary realm of the laws of war. When we retreat to the familiar refrain of "it's legal," we neglect the larger and far more important realm of the moral and ethical. We cannot be satisfied with this retreat, comforting as it tends to be. The lawyers do not have the answers to the questions we struggle with, and in seeking only their advice we might just rob ourselves of the kind of thinking required to answer some of this generation's greatest quandaries.

In a lecture at the Naval Academy, Michael Ignatieff spoke about these "habits of the mind" we have in always seeking legal review. He says this encourages "the view . . . that if you have legal coverage, you have moral coverage," but "what is legal is not necessarily moral." He goes on to say if someone signs off on your plan,

> you have to execute, but do not fool yourself. The moral debate inside you is not over. A moral service and ethical service is a service in which every person takes upon themselves the moral responsibility to ask: am I comfortable all the way down with this kind of stuff? And when we take the ethical decisions, and we hand them to someone else, we can begin a process of moral abdication. Ethical life is too important to leave to lawyers.[32]

Will the privates read Kant? Perhaps not, but—in matters of conscience, in matters of national policy, in matters of war—the law, though necessary, is simply not enough.

4

The Spectra of Impunity in Warfare

At that time I really thought that it would abolish war, because of its unlimited destructiveness and exclusion of the personal element of combat.

—Nikola Tesla[1]

When Nikola Tesla made the statement above, he was talking about his proposed invention for a directed-energy weapon offered to defense establishments in 1898 and 1900. Though it was never built, and it is unclear whether Tesla could have come through with the promise of the finished product, there is no reason to believe his hope of ending war as a result of its fielding is anything other than the kind of wishful thinking humankind has practiced on the subject throughout history. Tesla died in 1943, but not before predicting the form of warfare in which the United States has been engaged in such places as Yemen, Iraq, Afghanistan, and Pakistan. He said, "In an imperfect manner it is practicable, with the existing wireless plants, to launch an airplane, have it follow a certain approximate course, and perform some operation at a distance of many hundreds of miles. A machine of this kind can also be mechanically controlled in several ways and I have no doubt that it may prove of some usefulness in war."[2] Tesla wrote those words in 1919.

On November 3, 2003, a Predator unmanned aircraft launched a Hellfire missile and killed Abu Ali al-Harithi in Yemen.[3] It was the first attack of its kind by the United States since President Gerald R. Ford outlawed assassination by Executive Order 11905 in 1976.[4] It was not Tesla's death ray, but it might as well have been. The U.S. defense and intelligence establishment had certainly found a way to prove his wirelessly controlled aircraft concept useful in war. The age of distant warfare—more accurately, warfare with impunity—had made a quantum leap.

The Oldest Dream of the Second-Oldest Profession

Ideas about killing from ever greater distance and with ever increasing force are as old as warfare itself, as is the penchant for warriors in every age to invent ways of protecting themselves. These are the evolutionary stages of the weapons and counterweapons of war. In hand-to-hand combat, the one with the longer arms has a distinct advantage. A longer bladed weapon changed this dynamic, as did a spiked ball on a chain or an even longer lance.

Around 1800 BC, approximately the same time the light chariot came on the scene, an age-old hunting and warfare weapon was improved to the point of domination that lasted for thousands of years.[5] The compound bow, made of wood, sinew, and horn, was capable of sending its projectiles hundreds of yards downrange.[6] An archer was then able to stand behind the blood and sweat of face-to-face combat and lob arrows onto the opposing side. If his bow was capable of greater distance than his rival's, he was effectively immune from attack unless he allowed his adversary too close an approach. Of course his was not a life without risk. Battle lines morph; strategic, operational, and tactical errors are made. The fog of war allows for its friction to chew up armies and relegate them to the pages of history. Knowing this, and in an attempt to live to write that history, the warrior has ensured that self-protection is also always a part of war.

Nonetheless, past efforts at regulating implements of war that reduce the risk for a set of combatants are quite common. In 1139 the Second Lateran Council declared the crossbow a "disgraceful" weapon because it "could be used from a distance thus enabling a man to strike without the risk of himself being struck."[7] It was considered "dastardly" for its ability to unhorse a knight or turn the tables on warriors of that other holy war, the Crusades. Of course the council thought it only disgraceful when used by the hand of infidels against Christians. Vice versa, it was a merry weapon of war, and its use was encouraged.[8] Such is the jaded eye of those engaged in the immortal contest of war; such is the inequity in the standards of conduct, often written by the stronger side. The latter is telling, for there is responsibility there.

Weapons and countermeasures that increase lethality for one in relation to the other—or put another way, reduce the risk to one while holding another at higher risk—always cause perturbations in the thinking on what is right, wrong, and legal in war. This is the argument for continuing to increase our technological edge and farther remove our soldiers from the risk of death in battle. It is the argument Senator Warner made in the 2011 NDAA. It is the argument often referenced in DOD unmanned systems "roadmaps." From a limited view of morality (perhaps not quite as narrow as the Lateran Council's), many claim a moral imperative to protect our soldiers from harm. They claim this is the vector of technology, and as in the past when norms of behavior changed to catch up with the technology of war, so they will again. We'll see. For now I will offer a slightly different view of the spectrum of impunity in warfare as it relates to technological trends.

Faster, Farther, Longer: Technology Trends in Warfare

Martin van Creveld's book *Technology and War: From 2000 B.C. to the Present* is a masterwork detailing the evolution of technologies throughout the history of warfare. Through the ages others have argued for various "revolutions of military affairs" coinciding with major technological leaps. The intent here is not to argue over revolutions in military affairs; I leave it for historians to ponder and debate. What seems clear is that three significant trends in warfare have accompanied the general growth in technology. What is perhaps less clear is the effect these trends have on what I will call the spectra of impunity in warfare.

Technology tends to *increase the speed of warfare*, both in the movement of combatants to and in battle, and in the time from strategy to execution or individual decisions to action. It tends to *increase the distance weapons can be used effectively* and therefore the numbers of targets that can be held at risk. Finally, technology has tended to *increase combatants' ability to make battle persist.* Interestingly, while the trend in technologically advanced countries has been toward greater impunity for the combatant and noncombatant, the same does not hold for those in less-developed states. Technological inferiority, which I use simply as a relative term, tends to result in populations that are geographically coincident with the war-making entities. It means they have to live in the cross fire of battle when technologically superior nations' populations generally do not, certainly not to the same extent as their technologically inferior adversaries. Additionally, there is a paradox at the vortex of these three trends, beginning with remotely controlled unmanned warfare and leading into global nuclear war where the impunity we desire is rapidly reduced to the detriment of *all* noncombatants, regardless of technological capability.

Over the course of centuries, the speed of warfare has increased, both in the implements and decisions of war. Oxcarts gave way to the faster chariots. Infantry went from walking to riding on horseback to riding in trucks and armored personnel carriers to being flown into battle in helicopters or troop-carrying fixed-wing aircraft. It took Napoleon three months to march to Moscow in 1812.[9] Covering roughly the same distance as Napoleon, U.S. forces took only twenty-one days from stepping off in Kuwait to pulling down Saddam's statue in Firdos Square in downtown Baghdad.[10] Similar increases in speed accompany technological advances in all domains. Sailing ships lost out to steamships, which gave way to diesel ships. Today nuclear-powered vessels travel at classified speeds exceeding thirty-five knots. Aircraft have gone from the nearly walking speed of the Wright Flyer to speeds in excess of three times the speed of sound.

But physical speed is not the only thing that has changed. The decision process and a leader's ability to turn strategy into execution or a tactical decision into action have sped up with the improvement in communication and information-processing technology. Consider the sleepless nights of King George III as he awaited word from his far-away generals during the initial tectonic fracturing of the British Empire. He could receive news from the theater of war only by ship. On his opposing side was the ride of Paul Revere: one lantern if the British Army approached by land, two if by sea. At Gettysburg, Robert E. Lee's ability to manage the battle was crippled by his long external lines, while George Meade enjoyed tight internal lines allowing him to personally converse with his generals the night before the fateful battle. That all changed with the eventual use of wireless radio and satellite communications. These advances meant strategic and tactical leaders could get information faster than ever before, which drove the need to increase the speed of the decision cycle.

Former Air Force fighter pilot John Boyd was known, though not widely enough, for many things—not the least of which was being a thorn in the side of anyone whose opinion he felt was erroneous. He created the theory of energy maneuverability and his famous "EM diagram" by which one can tell when two opposing aircraft have the aerodynamic advantage at various energy states. No aeronautical engineering student gets through school without understanding specific excess energy—the heart of the EM diagram—and no young fighter pilot gets through training without comparing EM diagrams of his aircraft against all those he dreams of one day shooting down. Boyd changed not only the way aircraft were flown but also how they were designed.

The EM diagram was not Boyd's most famous creation though. He is the inventor of the "OODA Loop." This is a decision model whose letters stand for "observe," "orient," "decide," and "act,"[11] and it has been used in the military and in business as a way of staying ahead in the decision cycle. The commander or executive who can figure out a situation, decide how to proceed, and execute that decision will normally keep his opponent in a reactive state that eventually leads to the disintegration of the opponent's ability to compete. The idea is called "getting inside the adversary's OODA Loop." Information technology, coupled with other military technology and an intellectual ability to use it, allows a decision maker to remain ahead in the decision cycle—inside the OODA Loop—while others who cannot compete fall by the wayside.

Today communication occurs at the speed of light; the decision cycle has to keep up, but there are those who claim it cannot. One argument for autonomous systems is the speed of future decision cycles outpacing the ability of the human mind. In essence, there is a human limit to the "size" of the OODA Loop. There are those who claim military systems will be "too fast, too small, too numerous, and will create an environment too complex for humans to direct."[12] Perhaps—if we allow it. Regardless, technology speeds all aspects of making war, from strategy all the way to the individual actions of soldiers on the battlefield.

It also increases the size of that battlefield. Technological advances throughout the history of warfare have trended toward weapons able to employ at increasing ranges. The composite bow and the "dastardly" crossbow are but two early examples. Catapults enabled practitioners of siege warfare to heave diseased carcasses of animals over city walls, an early form of biological warfare. Rifles and cannon put rear areas under fire. Mortars are lobbed onto air bases in the middle of Afghanistan from kilometers away. Aircraft increased the reach of naval power and eventually allowed ships that never saw each other to duke it out over hundreds of miles of open ocean in such epic battles as the Coral Sea and Midway. Tomahawk missiles fly hundreds of miles to their targets. Predator and Reaper operators sit in air-conditioned trailers in the Nevada desert and launch missiles from remotely piloted aircraft thousands of miles away. And humankind's ultimate weapon (so far) sits atop intercontinental ballistic missiles (ICBM) able to cover the globe in thirty minutes or less and unleash the heat and light of tens of thousands of suns upon all the inhabitants of the earth. In a perverse consequence of the struggle for impunity, we now have a technology capable of destroying ourselves. We have pinned the hopes of our species on the idea that in holding every being at risk of annihilation, we will be forced to live in peace.

Those technologies that allow weapon employment at greater ranges are enablers of the distance trend. These include onboard navigation systems, offboard navigation satellites, communications, propulsion, seeker technologies specializing in various portions of the spectrum of light, and so forth. Many of these have aided the move toward greater precision in weapon delivery, and it is this precision, perhaps most of all, that allows greater effective ranges. There was a time when men fought to stick knives in each others' bodies or deliver blows with balls on chains. These were brutal but very precise means of accomplishing the dirty task of war. Longbows and crossbows allowed this same precision from greater range, and this is exactly what we see today with precision ordnance of all kinds able to be employed from distances that were only the stuff of dreams for longbow archers. Precision ordnance is not a trend in and of itself, and it is probably not a revolution either. Accuracy and range have evolved together over many centuries. Both are intimately linked to warfare with impunity, but in the end they are enablers of the trend of distance.

Technology in warfare also expands war makers' abilities to fight in the dimension of time. It allows for the persistence of battle. Napoleon lost his army during the cold return from Moscow. Summer was the season for fighting in Russia. America's Revolutionary and Civil Wars were mostly daytime affairs. It is hard to see what to shoot in the black of night. Though there is talk of the "fighting season" in Afghanistan, we now have the ability to fight whenever we choose. Night-vision goggles would have been a disruptive technology during Washington's trip across the Delaware River. During Operation Iraqi Freedom, satellite-aided bombs fell through huge sandstorms that would, in centuries past, have signaled a de facto truce in order for both sides to hunker down and survive. Some of Saddam's forces were not so lucky. Technology allows for the persistence of battle, from the intelligence gathering that feeds it to the actual engagements of war. It is at the convergence of these three trends where a view of the spectra of impunity comes into focus.

The Spectra of Impunity and the Paradox of the Technology Trend Vortex

The speed of war, the distance at which we can strike, and the persistence with which we can do so make up the trends in the technological advancement of warfare. With each of these, for belligerents who could harness them, there has been a tendency toward increasing impunity and reduction of risk to combatants. There are exceptions. In a dearth of intellectual ability that failed to capitalize on the promise of the armored tank, hundreds of thousands of Europe's youths were cut down in their prime when the stagnation of trench warfare met the automatic fire of the machine gun. Still, in general, growing technology means a gradual move away from the one-on-one distance of hand-to-hand fighting. Additionally, even a slight technological advantage, or more accurately a better mastery of technological trends, resulted in lower risk to that side.

One need only look at the data for the Gulf War and later conflicts to understand how superior technology, primarily in conventional warfare, affects risks to combatants. In Vietnam there were 127,405 casualties for a casualty rate of .03 percent per day as a percentage of daily theater strength (240,000). In the Gulf War there were 366 casualties at a rate of .0016 percent per day as a percentage of daily theater strength (530,000).[13] Then in the air war over Kosovo, NATO casualties reached 0 percent. Precision enables greater distance and impunity for the technologically superior. Belligerents who are able to harness the technology trends find decreasing casualty and combat death rates.

Along the technological spectrum, the risk to noncombatants seems to vary greatly from combatant risk. It also varies with the side the noncombatant is affiliated with in relation to the technology trends and where the noncombatant resides. As the speed and distance of war increases, a greater percentage of the civilian population unlucky enough to be geographically located in the war zone is held under the gun. In the history of war, there came a point very early on when armies began to reap the spoils of war on the backs of innocent civilians as they plundered, pillaged, and committed other crimes during campaigning. Here the civilian population was put at risk almost entirely due to geographic coincidence with fighting. However, as technology trended toward faster, farther, and longer, more geocoincident populations or those of technologically inferior belligerents found themselves at risk from the *direct* effects of war. They suffered now not due to armies taking liberties outside of combat but actually because of combat. The bombed-out ruins of Coventry Cathedral and the devastation in such cities as Hamburg in the last world war are illustrative of this trend.

On the spectrum of impunity, there comes a point where noncombatants are held at greater risk than technologically superior combatants. It could be no clearer than the example of Operation Allied Force over Kosovo. Although there were no U.S. or coalition combat fatalities, it would be pure ignorance to believe there were no noncombatant deaths. In 2000, Human Rights Watch reported that approximately five hundred civilian deaths occurred during the operation; the then Yugoslav government claimed twelve hundred to five thousand.[14] Anything greater than zero means noncombatants in the former Yugoslavia were at greater risk both because of NATO's ability to use technological trends to its advantage and because they happened to live in what turned into a war zone. This risk to geocoincident noncombatants and those of technologically inferior nations or actors is exacerbated in unmanned or robotic warfare.

Unmanned systems carrying out kinetic strikes, the most prevalent of which are currently in the form of targeted killings, lower the bar for combatant risk even farther than what NATO aircrews experienced in Kosovo. Here there is no human in the attacking weapon system, and its operator is not even in the theater of operation. Again, even one noncombatant death is evidence of greater noncombatant risk, but the data really are murky due to the nature of the operations. The Special Rapporteur on Targeted Killings and Extrajudicial and Summary Executions to the UN Human Rights Council stated:

> States have also refused to provide factual information about who has been targeted under their policies and with what outcome, *including whether innocent civilians have been collaterally killed or injured* [emphasis added]. In some instances, targeted killings take place in easily accessible urban areas, and human rights monitors and civil society are able to document the outcome. In others, because of remoteness or security concerns, it has been impossible for independent observers and the international community to judge whether killings were lawful or not.[15]

In 2009 a *New York Times* article claimed in the three years prior there were fourteen targeted killings by drones in Pakistan, with seven hundred civilian casualties.[16] Israeli targeted killings, mostly occurring in the West Bank and carried out by various means including drones, helicopters, snipers, and artillery, accounted for 234 "target" deaths and 387 civilian deaths.[17] While these data are not pristine due to inclusion of other tactics and weapon systems, the trend seems to indicate during killings of this kind it is hard not to inflict some degree of collateral damage. Again, with unmanned systems and near-zero risk of casualty, the risk to noncombatants on the adversary side seems to be higher than that of the "combatants."

So what of the noncombatants fortunate enough to be geographically secluded and/or on the technologically superior side? For many decades, Western militaries have been able to insulate their civilian populations from the direct threat of the effects of combat by deploying to the hot zone and fighting in someone else's backyard. Even European militaries have been able to do this since the end of World War II. U.S. citizens have not felt the large-scale effects of war on their territory since the middle of the nineteenth century and the Civil War. While much of this is due to geography, in the modern era the ability to deploy across the world with a fighting force capable of invading and occupying entire states is unique to Western, primarily NATO, nations. This is because of the convergence of the three technological trends and those nations' abilities in effectively managing them. Because of this, noncombatant citizens of technologically superior nations have experienced near-zero risk of casualty.

I postulate, however, this will not continue to be the case. As recent mass-casualty terrorism indicates, actors without the ability to militarily attack and overcome technologically superior nations tend to find other ways to lash out. They are generally, at least to the present, not bound by similar views on the laws of war and the immunity of noncombatants.[18] It is conceivable in the not too distant future, victims of unmanned warfare might attack civilian populations in those nations waging such warfare. It is equally possible they might attack the operators of the unmanned systems themselves, either in the performance of their military duties or during off-duty time in cities as far from the "war zone" as the operators are when they prosecute their targets half a world away.[19] What will we say about that act and the collateral damage it is sure to bring? I wonder.

At any rate it seems logical to project, as unmanned and possibly robotic systems lower the risk to the "combatants" using them, the risk to their own noncombatants is likely to rise, perhaps above the level of the risk to the systems operators themselves. Were this to occur, it would be the first time in history that those charged with fighting and winning wars, those individuals who volunteered to shoulder such a weighty responsibility, would experience lower personal risk than the combatants they are charged with killing and the noncombatants on *both* sides. This is the fracture in the continuum where the long-standing rules of war begin to crumble. This is the point beyond which we have no knowledge of how to exist in any meaningful state of war. This is *the* coming revolution in military affairs. And in the ultimate paradox of the technology trend vortex, our understanding is restored only

Figure 4.1 Spectra of Impunity

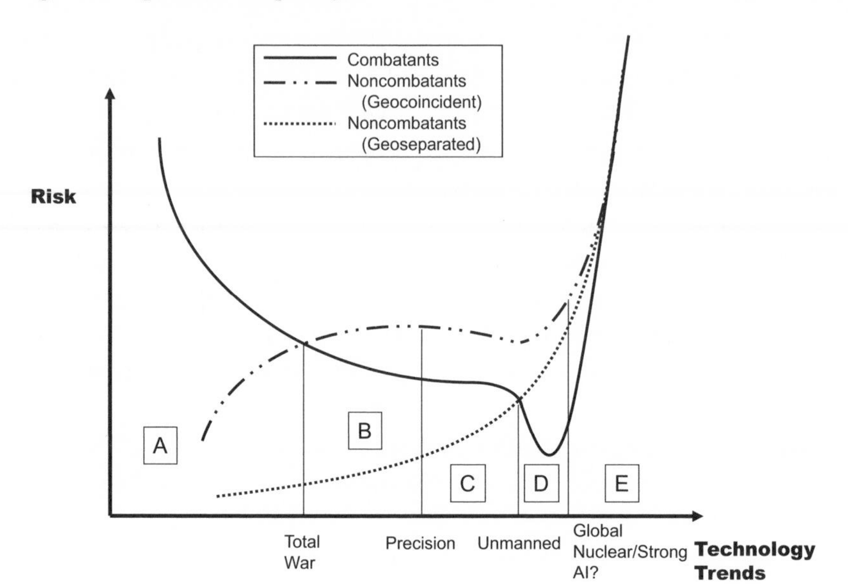

Illustration by author.

with the resort to all-out nuclear war where combatant and noncombatant are at least equals in the risk of annihilation. I refer the reader to the graphic depiction of the spectra of impunity in Figure 4.1.

Note first of all that this is simply a conceptual model of risk charted in relation to technological trends and the kind of warfare they allow. For instance, the advent of blitzkrieg and its counterpart in the Allies' "strategic bombing" allowed war as total as the world has yet known. Similarly, precision weapons, unmanned warfare, and nuclear capability drive differences in the form of war and the impunity enjoyed by those able to harness these enablers of technological trends. It highlights the spectra of impunity in relation to the drivers of technological trends and does not represent any quantitative representation of a particular data set.

The solid line represents combatants of technologically superior forces. The dashed and dotted line represents noncombatants of technologically inferior states or actors and/or populations that reside in war zones; they are geographically coincident with combat operations areas. The dotted line represents noncombatant citizens of technologically superior states or actors and/or populations fortunate enough to be geographically dislocated from the combat operations areas. Increasing technological capability and the trends it drives are plotted on the horizontal axis against the level of risk to life and limb to combatants and noncombatants. Major change points on the curves are highlighted by the vertical lines labeled beneath, and this divides the model into five different regions labeled A through E.

As technology trends toward the faster, farther, and longer in regions A through C, the personal risk to the combatant steadily decreases. As discussed above, very early in the history of warfare and therefore on the left side of the technology scale, the geocoincident population began to bear some risk of pillage, plunder, and armies "enjoying" what were then seen as the spoils of war. At the same time, and all the way through region C (the end of which represents the advent of armed unmanned systems), geoseparated populations or those noncombatants from the technologically superior side experienced very little personal risk (unless, of course, they became combatants by choice or conscription and moved onto that spectrum). Eventually geocoincident, technologically inferior noncombatants began to experience greater risk than the combatants due to the *direct effects of combat.* Tokyo, Dresden, even Hiroshima and Nagasaki, as well as World War II sieges of Soviet cities are examples, as civilian deaths tended to outpace technologically superior combatant deaths. This is represented at the boundary of region A and B, where the world experienced as close as it has ever come to total war. Interestingly, around this time there was a divergence in accuracy and distance as bombers were able to deliver tons of ordnance over significant ranges but without true precision, even though it was called daylight precision bombing. At some point the advent of precision ordnance began a reconvergence of distance and accuracy, and noncombatant risk lowered due to the ability of single weapons to pinpoint individual targets. Along with accuracy, there is an increase in destructive capability per round, per pound of ordnance, and per sortie flown,[20] but the population as a whole sees a reduction in risk of casualty. It is debatable whether collateral damage in the seventy-eight-day war over Kosovo was disproportionate, but the loss of five hundred noncombatants, or even twelve hundred to five thousand, is significantly less than in previous wars. Citizens of Baghdad during the Gulf War quickly understood they were not the targets and lost much of their fear of air attack due to the better precision capability of coalition air forces.[21] Still, it may be that up to 90 percent of casualties in modern war are noncombatants.[22]

At the right end of region C, combatant risk begins a significant downturn. This is the beginning of risk aversion, discussed in the next section and manifested in pre-war decisions on force structure and campaign type, such as the character of operations in the Iraqi no-fly zones and Kosovo, and in the decision to pursue and deploy armed unmanned systems. This leads to a situation where the relative risk of combat operations is transferred from combatants to noncombatants in a "risk inversion," which rattles the understood norms of the rules of war. This risk inversion is the postulated revolution in the structure and meaning of war aided by the development of robotics in warfare. It destroys what we know about how to interact in the depths of war because of the moral implications of risk discussed in the next section. Understanding this impact requires discussing the far right side of the graph first.

Prior to the era of risk aversion and the fielding of armed unmanned systems, the moral, ethical, and legal systems in place to restrain war and constrain its practitioners held together the concept of war as a meaningful activity: a just war is one that we are able to fight.[23] Then the world changed with the first successful test of a nuclear device in the New Mexico desert on July 16, 1945.[24] The United States is the only nation or entity, so far, to use a nuclear device "in anger." It did so in a time of extreme asymmetry in technology and—in what authors Gar Alperovitz and Sanho Tree call an "American myth"[25]—reduced its own combatants' exposure to the risk of invasion by destroying two Japanese cities. We will never know whether this supposition was actually true. Regardless, the bombings might have irreparably altered the spectra of impunity, but ultimately U.S. asymmetry was unsustainable. Eventually the major postwar powers gained the status of mutually existential threats. The world could no longer fight a war with its latest weaponry for fear of extinction of the species, and such a war's justice would surely be questioned. Mutually Assured Destruction as the ultimate form of deterrence has worked so far because of the rapid reduction of impunity, across the board, brought on by the prospect of a worldwide nuclear exchange. It turns out nuclear war is the great equalizer. It is this concept that challenges the simpler model of risk reduction purely through distance—what some have called distant warfare.

I have argued that technological trends have not simply moved combatants farther from each other and allowed killing at greater range. I have claimed the important concept is the relative impunity the three technological trends drive. Further, I will argue an imbalance in the spectra of impunity has an insidious effect on the resort to war and accepted norms of behavior in war. For those thinking only in terms of technology's ability to create distance in warfare, the quest—and argument—usually ends at ICBMs. This wholly misses the point. Nuclear missile warriors are not removed from the battlefield. They operate in it every day.

The missileer in a hole on the northern tier of the continental United States is no less immune from the consequences of his action than those in the cities his missile will find in less than half an hour. His location is on the other end of the trajectory of an ICBM launched from another belligerent state. The action of turning his key, when accompanied by the simultaneous turning of his partner's key, sets off a chain of events in which massive retaliation is the rationale in the first place or the probable end. It is at once comprehensible and inconceivable. One can understand the logic chain, yet its end result may mean the end of the species. It is hard to imagine taking such a chance, both for the legitimate authority charged with giving the order and for the key turner. The disproportionate nature of his singular risk in relation to all those he would kill is a powerful constraint, but it is the removal of impunity for all humankind that is more powerful still. Balance returns only with equal but near-zero impunity. The madness is held in check by the threat of extinction; it is a war that cannot be fought due to the just war principles of proportionality and probability of success. As the 1980s movie *WarGames* says, in thermonuclear war the only way to win is not to play.

This calculus is twisted in the risk inversion of region D. Unmanned warfare makes it possible for the first time for noncombatants on *all* sides to experience greater risk than technologically superior combatants. This undermines combatant moral status and affects the legitimacy of these weapons of war. Add to this the insidious rise in risk to noncombatants—we have not yet acknowledged it exists—and it impacts issues of last resort, the idea that all reasonable measures short of war must be exhausted before commencing hostilities, because we are blind to the consequences of our actions. "War," in quotes because it is unclear whether it qualifies anymore, appears too easy. The farther we proceed in time away from the last mass-casualty attack on our own noncombatants and the longer we go without some form of symmetrical response—al Qaeda drones over LA?—the less likely we are to recognize the imbalance in the spectra of impunity or properly deal with its moral implications. We are now entering our second decade since the first U.S. foray into unmanned killing. The Soviets restored the nuclear balance in less than half that time. It is past time we consider the moral implications of risk and impunity in warfare.

What Is There to Say about the Morality of Risk?

As I sat in the cockpit on the day I was to lead my squadron to war, my vice wing commander climbed the ladder to shake my hand and to leave me with an enduring thought. He said, "Bring them all home." Few other professions embrace the underlying culture of that statement. Though it is far more likely for a lawyer, doctor, or construction worker to be killed on the highway on the way to or from work than it is to die in combat in the twenty-first century, those events are the random acts of fate. They are not integral to the work those individuals do. Unlike general society, soldiers have no expectation of safety in the performance of their duty. They serve to put themselves in harm's way so others do not have to; in fact the very idea of service demands it. My vice commander did not wish me safe travels, for that would have been appealing to the whims of fate. What he said was to bring them home. It was not an order, merely a sentiment of the sacred responsibility of command. What he meant was left unsaid. Bring them home if you can; risk them if you must. It is a simple concept leading to the most difficult decisions humans can ever make. It is, however, an enduring part of the profession of arms. As Martin Cook says, "the military contract obliges military personnel to run grave risks and to engage in morally and personally difficult actions."[26]

Risk has always been resident in warfare. It comes in many forms: individual risk to life and limb, financial risk to the state, political risk to governments, risk of changes in the social contract, and surely others. Here, though, I will concentrate on the warrior's risk and leaders' decisions revolving around it. I will attempt to discuss the morality of this risk and its relationship to the just war tradition, for better or worse the ethic we use to justify our resort to and conduct in war. This discussion is the foundation for future thoughts on how our changing way of warfare affects the warrior's relationship to his counterpart and the subsequent effect of the very meaning of war.

Just war theory is broken into three distinct areas. They are concerned with resorting to war, conduct in war, and finally conduct following war and the return to peace. These are referred to as *jus ad bellum*, *jus in bello*, and *jus post bellum*, respectively.[27] Only the first two have bearing on the issues at hand. The areas of just war theory and their status as distinct conceptual frameworks is the essential crux of their validity in that, particularly in *jus ad bellum* and *jus in bello*, the responsible parties who decide for war—the legitimate authority—and those conducting warfare are independent from each other.[28] Their cause and actions can and must be judged separately. From this concept flow the key principles of distinction and noncombatant immunity, the moral equivalence and innocence of combatants, and the concept of accountability that form the vertebrae of the just war backbone.

There are four principles of just war theory deeply affected by the warrior's risk—two from the area concerned with the resort to war and two from the area of conduct in war. The amount of risk the legitimate authorities are willing to accept has moral implications for the probability of success and for proportionality, the latter in a manner not usually discussed. The individual warrior's risk in the conduct of war is integral to the rights of combatants. It is based on the concept of consent and the moral equality it produces, as well as the moral responsibility entwined in the doctrine of double effect, which sometimes produces the justifiable dissolution of noncombatant immunity.

For a war to be judged as just, it must have a reasonable chance of success. This is to prevent mass violence[29] or cause "death and destruction to no purpose."[30] Questions on this criterion might, for example, take the form of whether it is possible to defeat global terrorism or whether operations in Afghanistan can result in a sovereign government in Kabul when history would put very low odds on the chances. Just wars must also meet the proportionality criterion.[31] In essence, the costs must be worth that which is to be gained by resorting to war. Importantly, this is the *universal* cost, not just the cost to one side. This sets up what Michael Walzer calls the "radical responsibility"—the idea that political and military leaders are responsible for the well-being of their own people but also for the well-being of innocents on the other side. He goes on to say:

> Its proponents set themselves against those who will not think realistically about the defense of the country they live in and also against those who refuse to recognize the humanity of their opponents. They insist that there are things that it is morally impermissible to do even to the enemy. They also insist, however, that fighting itself cannot be morally impermissible. A just war is meant to be, and has to be, a war that it is possible to fight.[32]

I will return to this radical responsibility in later chapters and discuss its relationship to warfare with impunity and to autonomous weapons.

Much has been made in recent years about a perceived risk aversion in the U.S. military, its leaders, and the public they defend. What began in the aftermath of Vietnam was reinforced in the post–Gulf War years as "the hundred-hour war" and "the Revolution in Military Affairs" became catchphrases in the defense establishment and those who commented on it. The perceived revolution of precision weapons and the overwhelming might of U.S. air power, it was believed, had changed the nature of warfare, making it more surgical, more destructive for the firepower employed, and therefore less likely U.S. servicemen and women, and those of coalition nations, would be required to risk themselves in battle. The hundred-hour war—a moniker that dismisses the preceding month-long air battle—supposedly proved the argument and further solidified the public expectation of low casualties in future wars. But the risk-aversion rhetoric went into overdrive following Operation Allied Force, the seventy-eight-day war in the Balkans. Its result was what defense analyst Jeffrey Record dubbed "force protection fetishism" in a venomous article over the impracticality of this perceived culture and its relationship to the desired end state. He said, "Effective use of force rests on recognition of the intimate relationship between military means and political ends. Obsession with keeping the former out of harm's way, even at the expense of aborting attainment of the latter, violates war's very essence as an act of policy." [33] His is an argument about the classical military theories of Carl von Clausewitz. It is actually a moral argument at its core.

The most common critique of NATO's air campaign during Allied Force in the realm of risk is that pilots were not allowed to fly low enough to increase the accuracy of their weapons and create the kinds of effects that would have brought Serbian leader Slobodan Milošević to his knees earlier.[34] The refrain is that pilots were kept from flying at altitudes that would have increased the effectiveness of their weapons, and their inability to correctly prosecute the war made it longer than it should have ever been. The ghosts of Vietnam abound. The problem is this argument is false. It is based on an understanding of the physics of warfare that has not caught up with the guidance of modern precision weaponry.[35] It is simply not true that pilots were always held at high altitude. They were allowed to deviate in order to ensure proper weapon effects (for nonprecision ordnance or for weather in most cases) against certain targets. However, this critique of risk is, quite frankly, irrelevant. The issue with risk in Allied Force was much more basic. The campaign plan was not suited to the stated objectives, particularly the goal of ending the ethnic cleansing of Kosovo.

The moral issue of a perceived or actual risk aversion during Allied Force that led to the national command authority never entertaining the idea of ground troops is that the probability of air power alone stopping the ethnic cleansing in Kosovo was virtually nil. More importantly, the converse was probable. Further, and potentially morally incriminating, such a consequence was foreseeable.[36] The fact is, much to the surprise of U.S. and NATO leaders, Milošević lasted longer than the duration of the original plan and, knowing there would be no ground troops in Kosovo, used the aerial bombardment as cover for redoubling his efforts at ousting Kosovar Albanians from what he saw as greater Serbia. Sole aerial efforts at *controlling*—the word choice is important—populations or militaries on the ground have not worked ever since the British first tried it in Iraq in the 1920s, right up through the United States and Britain trying it again in Iraq from 1991 to 2003.[37] Ethnic cleansing is a one-on-one, in-the-face, personal "conversation" between an oppressor and the oppressed. No fighter pilot flying at any altitude has the means to stop or even enter that conversation unless he is willing to kill both. And that, as we say, would be a "show-stopper." If it was avoidance of risk or aversion to casualties that kept ground forces out of the plan for Allied Force, then it was the paradox Record speaks of—a concept of force that has almost no chance of success. More subtly, it is one "worth" doing but not worth expending those who might have the only chance of achieving the goal. This leads me to a concept of proportionality not often discussed.

Proportionality in *jus ad bellum* requires the state (or the legitimate authority in the case of wars for self-determination) attempting to fight a just cause to ensure that the universal cost—the cost to both sides—is worth the objective. Most often when a war is critiqued as unjust, it is because too much force was used at too great a cost to the combatants on both sides and to civilians on the adversary's side. For all the rhetoric, it is the proportionality argument that lies at the heart of critiques of the decision to go to war in Iraq in 2003. The basic question, knowing what we knew then and thinking we knew more, is whether toppling a regime and throwing a country into chaos was worth the threat Saddam's Iraq might have posed. The cost of intervention to both sides was unlikely proportional to the threat of the Saddam regime remaining in power, particularly when the primary rationale for war—and hindsight is always perfect—was found to be incorrect. Again this is the argument on proportionality most often proffered—that of too much force—but it seems to me it is not the only argument.

Assuming just cause and all the other just war requirements are initially met, there is also an obligation under proportionality to use *all* the necessary force to accomplish the mission. If risk aversion leads an authority to rule out certain military options that may be required in order to accomplish the task, then the problem is not committing enough force to meet the ends. Proportional force must be worth the cost, but the decision to use too little force due to a misapplication of the principle to only a single side of the conflict without taking into account the universal nature of proportionality cannot be justified either. In effect it drives a regression to the probability-of-success argument stated above. It also blurs the line between *jus ad bellum* and *jus in bello* if the rationale for such action is a reduction of risk to one's combatants being taken at the expense of noncombatants.

Perhaps the most easily argued position for risk in the morality of war lies in the doctrine of double effect. This idea comes from a "principle of moral reasoning that argues the pursuit of a good end tends to be less acceptable when a resulting harm is directly intended rather than merely foreseen."[38] The argument goes as such: performing an act likely to have evil consequences—in this case the killing of innocents—is justified as long as the following four conditions hold:

1) The act is good in itself or at least indifferent, which means, for our purposes, that it is a legitimate act of war.
2) The direct effect is morally acceptable—the destruction of military supplies, for example, or the killing of enemy soldiers.
3) The intention of the actor is good, that is, he aims only at the acceptable effect; the evil effect is not one of his ends, nor is it a means to his ends.
4) The good effect is sufficiently good to compensate for allowing the evil effect [i.e. justifiable under proportionality].[39]

As Walzer explains, the burden of the argument is in the third clause, and for this reason the doctrine is sometimes referred to as the doctrine of double intent.[40] In military terms this means the killing of innocents is sometimes justifiable assuming legitimacy of previously held *jus in bello* requirements as long as the bearer of military force does not intend the deaths and the deaths do not serve to further his aims. The moral arguments do, however, become more stringent. Proportionality, for instance, is twice required—first in assessing the right use of force to achieve the objective and second in the case of double effect, that the objective is sufficiently good to allow for the killing of innocents. In the end the doctrine of double effect "is a way of reconciling the absolute prohibition against attacking noncombatants with the legitimate conduct of military activity."[41] It is a difficult concept to grasp and one that is even harder to be judged to have met.

What the doctrine of double effect means to those charged with conduct in war is that if there is a way to prevent or mitigate the poor consequence of killing innocents, we are morally bound to do so even if it requires greater risk to ourselves. Walzer puts it this way: "When it is our action that puts innocent people at risk, even if the action is justified, we are bound to do what we can to reduce those risks, even if this involves risks to our own soldiers."[42] To return to the Operation Allied Force example, the administration's unwillingness to consider using ground troops to intervene in what many saw as a civil war, along with the public statement confirming the decision, allowed Milošević to carry out more effective ethnic cleansing at an exorbitant cost to the Kosovar Albanians. We placed others at risk while refusing to accept a proportional amount for ourselves "even when that acceptance was necessary to help others."[43] Walzer further states in discussing the Kosovo issue (quoting Camus in italics),"Well, I have no love of battles, and I fully accept the obligation of democratically elected leaders to safeguard the lives of their own people, all of them. But this is not a moral position. *You can't kill unless you are prepared to die.* But they cannot claim, we cannot accept, that those lives are expendable, and these are not."[44] In concluding the thought in *Just and Unjust Wars*, he says, "A fixed policy that their lives are expendable while ours are not can't, it seems to me, be justified."[45] This is where risk aversion and robotic systems collide.

In what seems like understatement, an article in *CQ Researcher* says that "some critics are concerned about the moral implications of weapons that put civilians—but not the weapons' operators—at risk."[46] If drone strikes are directed against legal targets[47] and such strikes are likely to kill noncombatants, and if a lower collateral damage method (such as a sniper) is just as effective at killing the intended target, then it is likely incumbent on the executor of military force to use that method even if it entails greater risk to the combatants. Here we walk a fine line as Walzer alludes, for we are also bound to risk our forces only as required. This is the commander's quandary that my vice wing commander ensured I understood by giving me a firm handshake and thoughtful look straight into my eyes that early morning as I prepared to take a squadron to war. But this is also the balancing act leading to the American way of war that increasingly chooses a five-hundred- or two-thousand-pound bomb from thirty thousand feet or a Hellfire missile from seven thousand miles away—or possibly an autonomous lethal robot with human risk as an irrelevant concept—when a well-trained Marine with a rifle would have sufficed. This is the flip side of the technology trap where states with precision capability may be required to use them to reduce collateral damage.[48] In a conceivable future where technologically superior states are required to use only precision capabilities while those without such capabilities are not, the international community may only be reinforcing a tendency toward risk avoidance with subtle but dire consequences.

Michael Gross, writing in the context of the Israel Defense Force incursion in Lebanon in 2006 and Gaza in 2008, is blatantly clear on the *jus ad bellum/jus in bello* bridge created by the relationship between risk and the doctrine of double effect: "When a nation or group goes to war, it has to decide how much risk it is willing to shoulder to achieve its political aims. If a society is unwilling to risk the lives of its soldiers in armed conflict, then it must search for alternative means to realize its goals. While combating armies have an obligation to protect their soldiers to the greatest extent possible, no army can mitigate risk on the back of innocent noncombatants."[49] This argument, beginning here as a moral one, eventually leads to a much more practical one based on the paradox that our current concept of military capability may, in the end, not allow for the achievement of military objectives in the pursuit of national interests. It is Record's argument with a new twist. This matter for policymakers will be discussed in the chapter on our "radical responsibility," but I now turn to the final just war principle affected by the concept of risk: the moral equality and innocence of combatants.

Moral judgments on the legitimate authority charged with the decision to engage in war necessarily differ from judgments on those responsible for conduct in war. There is an independence—a "useful fiction," Gross calls it—between the cause and "fighting well." The individual soldier is exempted from judgment on just cause but held accountable for his or her actions in war.[50] It is possible both to commit war crimes in a just war and to fight justly in an unjust war. Consider Field Marshal Erwin Rommel, the German genius of armored warfare during World War II who ignored Hitler's orders to execute enemy soldiers caught behind the lines. He was a smart man and was surely aware of the uncertain nature of his cause, but "he fought a bad war well."[51] The idea that soldiers are judged apart from their cause sets up two important moral principles: the moral equality and innocence of combatants—innocent in the sense they are not judged as criminals for their legitimate acts of war.

In war, combatants consent to being killed when they step onto the battlefield, as long as they maintain their combatant status. Because the act of killing is less restrictive in the law of war than in criminal law, the status of combatants and noncombatants gives rise to certain rights. Combatants' consent to be killed grants rights on how and when they may be killed and protections against being killed when they are *hors de combat*, either unable or unwilling to continue fighting due to neutralizing injury or surrender. One of these rights is the right to be judged as a warrior, for the act of killing in war is not the same as the act of killing outside its limited constraints. In criminal law killing is prohibited unless justified. In war it is allowed unless prohibited, but the consent of the combatant to have the same happen to him means that his enemy cannot then be judged a criminal for the same act. In Walzer's words, "without the equal right to kill, war as a rule-governed activity would disappear."[52] Combatants are moral equals and innocent to live as free men when they lay down arms at the end of a war, assuming they "fought well." This we allow—indeed insist upon—because they are judged apart from the cause and because they consent to the equal treatment from their enemies on the battlefield. As Walzer says, "I find them my moral equals. That is not to say simply that I acknowledge their humanity, for it is not the recognition of fellow men that explains the rules of war; criminals are men too. It is precisely the recognition of men who are not criminals."[53]

Combatants live in the same chaotic world of life and death. They are, in a sense, inextricably bound to each other in a shared fate within the blinding light and deafening sound of men trying to kill and survive. But what happens when one no longer decides to risk the threat of the sort of life his opponent has to live? What happens when our "combatants" are no longer granting their consent to the consequences of the battlefield? It is probable they risk their moral equality and with it the rights granted under its umbrella. As Ignatieff explains, "the tacit contract of combat throughout the ages has always assumed a basic equality of moral risk: kill or be killed. . . . But this contract is void when one side begins killing with impunity."[54]

The consequences of a refusal to accept any risk on the battlefield are morally clear but legally ambiguous. Without consent to whatever may occur in battle, the combatant's moral standing as a combatant cannot exist. If he is unwilling to occupy the same world as his opponent, he has no right to expect to be judged as innocent for the killing he does. Murderers often kill out of a sense of moral superiority or self-gratification. They see their victims—a word not used in the language of war—as inferior or worth less than whatever emotional boost they receive from the act. Superiority complexes often lead to atrocity and humiliation, the Third Reich and its Holocaust being not-too-distant reminders. From author and professor of international relations Christopher Coker: "There is no point in humiliating an enemy, for the warrior lives in the same world of death and in the imagination of the other. Warriors form a guild, sometimes even a fraternity."[55]

At the same time, one who does not consent to the risk of war has no reason to see his opponent as innocent of the killing he does either. Neither sees the need to judge each other apart from their cause. They are likely to see each other as fighting an unjust cause while acting in a criminal manner. Perhaps this is why we seem to have no qualms about the targeted killings of "illegal combatants" outside of what some consider the legal definition of the combat zone[56] and saw no issue, at least institutionally, with denying basic Geneva protections[57] to detainees in cases such as those of extraordinary rendition. We do not seem to respect the moral equality of those we are killing in drone strikes, and it seems equally clear we are denying basic combatant rights to those who are *hors de combat*—otherwise we would call them "prisoners of war" and grant them the rights the classification demands. Unfortunately this is not some dreaded future that may or may not come to pass. Refusal to risk in warfare changes the calculus of consent and indelibly alters the moral foundation of what it means to be a combatant. We will not likely know its true consequences on war as a meaningful human activity for years to come, but we know we have changed it, perhaps forever.

Consent and Meaning in War

There is a spectrum of impunity in warfare, but it likely does not conform to the common view of hand-to-hand combat on one end and ICBM-delivered nuclear warheads on the other. It is far more nuanced, and it is why we have such difficulty quantifying it. We equate launching missiles from Predators to F-15Es dropping laser-guided bombs because we have not fully dealt with ambiguities of the spectra of risk and impunity in war. Scholars already question clandestine missiles landing in countries with which we are not at war,[58] but they have not adequately dealt with the issues surrounding the physical location of the combatants making lethal decisions, the effects of which are felt half a world away. They have not dealt with—none of us have—the risk inversion of the quest for impunity in unmanned and autonomous weapons and its implications.

To be sure, the warrior seeks safety for himself while increasing lethality against his foe. It is why, quite frankly, we fight as we do. Yet up to now there has always been some element of risk involved. Tomahawk shooters are afloat in often hostile seas. They are vulnerable to attack from sophisticated forces or those with a speedboat full of explosives. Snipers walk for miles in hostile territory and find a hole-up site affording them a clean shot while, at the same time, maintaining concealment. Fighter and bomber crews flying "risk free" at medium altitude over Kosovo dodged surface-to-air missiles for seventy-eight days; two were shot down. The same crews staged out of bases in Iraq and Afghanistan for operations in those two countries, and their bases were shelled on a daily basis with indirect fire from mortars. They knew they had little chance of survival if even a minor mechanical failure brought their aircraft down over insurgent-controlled areas. They knew unimagined atrocity would precede their inevitable, eventually welcomed death, but they admitted it to themselves only in times of weakness, such as during early morning goodbyes or when tucking kids in bed the night before deploying.

For those who claim operations as I have described are no different than firing Hellfires from Nevada or, in the future, sending in the "droids," I wonder how they would have taken to the opportunity to strap into a single-engine fighter—meaning your life literally depends on that power plant—and fly over improvised explosive device (IED) attacks, troops-in-contact situations, the massive slums of Sadr City where distressed aviators would be unlikely to find hospitality, or just the openness of a hostile and seemingly limitless desert . . . then sleep, eat, brief, and do it all again for hundreds of hours over several months. There is risk, even in our technologically advanced form of warfare, but there has never been a military or its civilian authority that has ever explicitly stated or implicitly acted as if it is unwilling to accept any risk, that it would rather kill from afar without ever exposing its soldiers even to the theater of war. We have not yet done so for war as a whole, but we certainly have for a segment of our weapons of war. We are only going farther in that direction, and as we do so we are failing to deal with the moral and practical implications.

Four principles of just war are adversely affected by the reduction of risk and our drive for impunity through unmanned and robotic warfare. A refusal to risk generates questions about the probability-of-success criterion and the importance of certain actions if the necessary force is not committed to achieve the objectives. This is the other side of the proportionality coin; authorities must commit the force necessary, otherwise they are simply wasting effort and lives. If these are noncombatants' deaths that might have prevented had a riskier solution not been bypassed, a lack of willingness to risk has caused unnecessary collateral damage that is unjustified under the doctrine of double effect. This is the moral hazard of the risk inversion on the spectra of impunity. Insulating combatants from risk, while exposing noncombatants on all sides to greater risk, is morally unacceptable.

Finally, it might not even be possible for those engaged in remote warfare to maintain combatant status and all the rights it affords. Fundamental to the rules of war is the consent of combatants to the possibility of suffering harm during legitimate acts of war. This is what creates the moral equality of combatants and eventually ensures their innocence and right to fair treatment when they are no longer able or have lost the will to continue the fight. Without combatant consent, the rules of war have no meaning, and the basis of just war theory is undermined, perhaps to the point of breaking. We must consider these issues if we are to keep war a meaningful activity and protect the rights of those engaged in it.

5

Impunity and the Politics of War

Remember, whatever virtue and goodliness there may be in this game of war, rightly played, there is none when you . . . play it with a multitude of small human pawns . . . [when you] urge your peasant millions into gladiatorial war.

—John Ruskin[1]

In reflecting on a life of service to the nation, two-time secretary of war Henry Stimson said, "As I look back over the five years of my service as Secretary of War, I see too many stern and heartrending decisions to be willing to pretend that war is anything else but what it is. The face of war is the face of death; death is an inevitable part of every order that a wartime leader gives."[2] It has always been so, and it remains so today. Yet there is a sense we are on the verge of changing the face of war.

On December 7, 2011, many Americans remembered the "day that will live in infamy," when Japan attacked Pearl Harbor, and the reluctant entrance of the United States into what became known as World War II. Four days later, as we busied ourselves with preparations for the holidays, what went nearly unnoticed was the seventieth anniversary of the last U.S. war declaration that led to hostilities.[3] It seems odd, seeing as how the United States has participated in at least eight large-scale armed conflicts since that fateful day. Three even lasted longer than that world war, and one has lasted twice as long. We seem to have had no qualms about military action over the past seventy years, engaging in combat operations roughly once per decade. Though our penchant for fighting seems unquenched, we tend to call our adventures anything but "war." We have had everything from police actions to, in vogue since the 1980s, "operations." America's longest war is called Operation Enduring Freedom. It certainly has endurance.

Dimensions of Impunity, Pooh, and the Poison of War

In all of these wars, Americans, their enemies, and many noncombatants have lost their lives. They have all known the "face of death," but the language we use to describe them, and our failure to declare them for what they are, are symptomatic of the vector of our way of war and the technologies that drive it. We like to hold the reality of war beyond our senses. We have wholeheartedly taken the advice of author A. A. Milne, the writer of the beloved Winnie the Pooh stories, when he called war poisonous, inherently so, and warned, "We should not roll it meditatively round the tongue and wonder how to improve the taste."[4]

Secretary Stimson surely wished his observation was not so. After all, we tend to shy from those things that cause us pain. It is the human thing to do, and yet there are consequences to those desires as well. Over the course of the last seventy years we have learned to fight wars with increasing impunity. We have held targets the world over at risk. The sun never sets on the U.S. military or on its ability to do harm to the enemies of the nation it serves. In building such a formidable force, we have done so with the expressed intent to bias the playing field, to protect our military members and the citizens of our country from harm, while dealing great destruction on those who harbor ill will toward us. We seek what Martin Cook has called "immaculate war"—war without casualties to our side.[5] Warfare with impunity is perhaps as close as we come to a national military ethic, and because it is so we continue along the technology trend vector without adequately considering the consequences of our actions. What have we done to restraint? Have we altered the psychology of war for both the citizen and the warrior? Have we driven the emergence of constant conflict and low-tech asymmetrical challenges to our perceived power? Is there even such a thing as war anymore? What about the warrior? Perhaps now is the time, as we stand on the fault line between warfare as a human activity and the world where it is not, to ask these basic questions.

In the previous chapter we dove into the theory of the risk inversion on the spectra of impunity and discussed just war problems with the reduction of risk in warfare. The risk inversion and its impact on just war, however, are not the full story. There are other dimensions of impunity shaping our understanding of warfare. These dimensions are intimately connected to *ad bellum* and *in bello* principles, but they differ from our earlier considerations in that they attempt to explain how impunity affects the psychology, ethos, and culture of all who have a part in warfare—its leaders; the citizens consenting to their governance; and the warriors, or at least combatants, on both sides.

The first of these dimensions of impunity deals with the politics of war. It is this dimension where we find the political barriers to war, the restraint the masses exert on their leaders. Impunity, or fear of reciprocity on the other hand, affects the psychology of the state's leaders—in just war terms, the legitimate authority—as well as the psychology of its citizens. Warfare with impunity tends to lower the barriers to war, making it easier for politicians to justify it (perhaps "excuse" might be a better term) and for citizens to buy into it. There is a sense, however, that their consent is driven more by alienation from the hazards of war and from the military they employ than from any rational consideration of interests or costs.

Impunity also impacts an area not often discussed in this context, that of the civil-military relationship. There are questions about whether it is the place of the civilian-controlled military to provide "should" advice as well as "how" advice when it comes to military operations. This blurs the lines of the aforementioned separate realms of accountability in *jus ad bellum* and *jus in bello*. It questions whether the weapons the military desires change the last-resort calculus, either purposefully or by accident, and turn military leaders into the caricature of movie lore where "an unused weapon is a useless weapon."[6]

The second dimension of impunity deals with the effect on the warrior of the technologically advanced military. Killing with impunity—often from very great distance and only through interaction with computer monitors, click wheels, and far-away sensors—changes the dynamic of war's most final of acts. The greater the impunity, the greater the filters on the sound, sight, smell, and fear of combat. Yet those who kill with impunity know they have done so, and their proximity to that knowledge and to the family they go home to every night has its own psychological cost. They live in two readily adjacent worlds; at least they hope they do, for another affect of the asymmetry of impunity is the continuing desire of the enemy to overcome it. Though they have not yet done so, the possibility always exists that our adversaries eventually will overcome the remote warrior's impunity and find ways to exact the heavy toll of war.

Additionally, the extent to which remote or autonomous warfare may change the warrior ethos and the existential nature of war is as yet unknown. Will unmanned warfare once and for all destroy any vestige of a warrior ethos in technologically advanced militaries? It may, but the impact of such a result will not be known for a very long time.

Finally, the effect of warfare with impunity on the future of war, particularly if it lies among those who lack the technological capability to pursue a similar way of war, must be considered in any real discussion on the dimensions of impunity in warfare. It is almost an extension of the dimension of the warrior. Those whom we prosecute with our advanced weapons are no less warriors. In fact, it is likely they have greater claim to the title than we do, but they occupy a realm of lesser technological capability and, as a result, know no impunity. They can neither relate to fighting with it nor understand those who use it against them. It highlights the disconnect between our technological warfare and the warrior culture, and it affects our "conversation" as moral equals.

The three dimensions of impunity attempt to explain how we approach the moral issues of risk in war and the dynamics between elected leaders, their constituents, and the professional military sworn to the defense of the structure of state. They compose a conceptual construct allowing us to delve into the psychology of decisions involved in fighting wars, and they provide a window to major obstacles in our ability to manage the technological, policy, and cultural issues involved in any form of warfare. They shine a light on the specific challenges posed by the entry of remote, unmanned, or robotic systems into that sphere. As Peter Singer recently pointed out in the *Journal of Military Ethics*, talking about emerging technologies and morality is hard to do.[7] Unfortunately we cannot retreat to the comfort of Pooh Corner and its lovable, half-shirted bear. The issues confronting us are a damn sight harder than the travails of a blustery day.

Impunity, Politics, and the Sacred Trust

Several years ago I accompanied my wife to her church's general assembly in Fort Worth, Texas. I am not a member and so attended none of the official functions; my job was simply to make sure we all arrived at various events on time. The big controversies of the year involved defining, or electing not to, the role of sexual orientation in the clergy and voting on a resolution declaring the war in Iraq unjustified. At a dinner, a Texas barbecue, I went through the traditional exercise of discovery for the people my wife introduces me to on what it is, exactly, I do. It starts with generalities and ends, grudgingly for me, with my chosen profession and experience in flying jet fighters.

This conversation, though, was a bit different. The air was charged with the excitement of democracy—the idea of taking a stand. The woman I was talking to asked my rank, then instead of the usual litany, asked if I agreed with the war. I simply answered politely that I do not discuss my personal opinions on such matters. I had the strong impression that by virtue of my being there, she did not know I was not a delegate, she assumed I agreed with the idea of the resolution (as did she), and she wanted to hear me say so.[8]

Without knowing it, what she wished for could actually sow the seeds of insurrection. She had not fully considered the consequences of her desire to hear, one way or the other, a professional military officer weigh in on matters of the state that are clearly within the purview of its leaders and the citizens who granted them that power. As an officer in the U.S. military, I am sworn to "well and faithfully *execute* the duties" of my office, and my lowly position does not entitle me to advise political leaders, except through the voting booth, on questions of "should" in military matters. It highlighted for me the incredible complexity of the dynamic between our elected leaders, those who put them there, and the sacred nature of civilian authority over the military. These are constantly at play in matters of war. As author and National Defense University professor Ken Moss points out in *Undeclared War and the Future of U.S. Foreign Policy*, "Only public opinion, professional military judgment, and presidential and congressional political judgment would serve as restraints,"[9] but it is possible our drive toward impunity is altering this dynamic in ways never intended. The ability to prosecute a war with impunity lowers the barriers to war and affects the psychology of the state's leaders, its citizens, and the sanctity of the civil-military relationship.

The Psychology of the Leaders

Michael Walzer warns us that sending troops to combat represents a moral threshold; it must be crossed "only with great reluctance and trepidation."[10] The greater political threshold may actually be sending troops into an area in the first place. Once there, ramping up to combat could require less deliberation, but this is perhaps semantics. Walzer says our political leaders must worry about it; we want them to worry about war. In fact "they have to *worry*, they can't calculate," because the values cannot be "expressed mathematically."[11] What is it that cannot be calculated? In warfare there is much that evades raw mathematics, but what he is getting at here is the human cost of war—more specifically, the human cost to one's own soldiers, because as we have discussed the risk to one's own noncombatants in a technologically superior society and prior to the risk inversion is assumed away. The idea of sending America's youth into harm's way, often to their deaths, serves as a powerful constraint on war. It has likely kept many presidents, and their generals, awake at night. It has made them age faster than they might otherwise have and driven them to decide, in rare instances, not to seek reelection. Additionally, the need to seek the consent of those who will fight constrains decisions to go to war, at least in democracies. Walzer says that this need "would surely limit the occasions of war, and if there were any chance at all of reciprocity from the other side, it would limit its means too."[12] This is why presidents always seem to ask, and military men attempt to quantify, how many will be lost before making their decisions.

As we have seen, the numbers of potential casualties was the stated reason for the decision to use atomic weapons on two cities in Japan during World War II. During planning for Desert Storm, Brig. Gen. Buster Glosson, charged with planning the air operation, briefed the president that he expected to lose no more than eighty coalition aircraft, probably closer to fifty (actual losses were forty-two).[13] Two days before the war, Lt. Gen. Chuck Horner, who, as commander of the Ninth Air Force and the U.S. Central Command Air Forces, was charged with executing the plan, estimated thirty-nine USAF aircraft losses (in fact the USAF lost fourteen) and no more than a hundred overall, including coalition aircraft.[14] Presidents are concerned with these numbers because the public is also concerned.

Public reaction to casualties, particularly when the objectives behind the use of military force are not widely supported, drive later presidential decisions on further uses of force. Mounting casualties in Vietnam, in a war that became less apparent as a required front to stem the tide of Soviet aggression, drove multiple changes to strategy and operational plans.[15] We have already discussed the perceived casualty aversion and seen how it might have driven operations over the Balkans and later confrontations. Just eleven days into Operation Iraqi Freedom—another war started without the "full" support of the public—56 percent of Americans polled said the United States "should do everything possible to minimize casualties," and 42 percent said the country was not using enough force against the adversary,[16] implying that with more force there would have been fewer American casualties. Public reaction and personal investment in the casualties of war certainly impact presidential decision making. Does a public reeling from the effects of dead and injured soldiers really mean the United States should do "everything possible to minimize casualties?" Has it really considered what such a strong statement means? As we pursue improvements in impunity, the obvious question is what happens if or when the constraint on considering the human cost—to your own side—no longer exists? As Moss puts it, "blood and death are not the only requirements for the human experience of war; the larger question is what the implications are when the architects of war find ways to use military force with as little human participation as possible. . . . Bloodless war seems highly unlikely, but what will be the implication for national leaders when or if they believe a relatively cost-free war in human terms is within reach?"[17]

Some believe there is an obvious answer. Martin Cook thinks new technologies allow for "military force with considerable accuracy and near-total impunity. Because of these technologies, political and military leaders can use military force in circumstances in which, in earlier times, they would not have."[18] For him the combination of precision weaponry and technologies that enable impunity have altered calculations on last-resort requirements and lowered the barriers to war. Speaking of Desert Storm and its impact on later conflicts, he says, "The downside of the Gulf War's precision, especially on the jus ad bellum dimension of just war theory, was made abundantly clear by the decision regarding how to use precision airpower in the Kosovo campaign. In that campaign the very precision of airpower, coupled with the impunity given by stealth and standoff weapons capability, served dramatically to lower the threshold for the use of military force."[19] Moss addresses the question on how presidents will react, with equal clarity: "When the prospects of human losses are low, presidents have found it easier to use force through their prerogative as commander-in-chief."[20]

There is ample evidence to back this up. President Bill Clinton, claiming "our target was terror," gave the go-ahead for the August 20, 1998, strike of seventy-five Tomahawk missiles on targets in Sudan and Afghanistan.[21] These were nations the United States was not at war with. Recall it had been fifty-seven years since a declaration of war, but these two nations shared other similarities. Both were implicated in support to Osama bin Laden, with one accused of manufacturing nerve gas at a claimed pharmaceutical factory. Both are large states. Both are landlocked. Both were suffering from impending state failure, though there is an argument that Afghanistan was already there. The failure of the Soviet Union in its ten-year war in Afghanistan and the recent memory of Americans dying in the streets of Mogadishu in another failed state must have weighed on the mind of the commander in chief. The monetary cost of landing a ground force in either of these states would have been significant, and the likelihood of more American boys (there were, as yet, no qualified female combat troops) dying in far-off lands for what might have been retribution or preemption was high. Cruise missiles seemed an easy choice, part of a simple cost-benefit analysis, and this example begins to illustrate a more subtle part of the psychology of political leaders.

Currently the combination of precision ordnance, the overwhelming strength of the American air forces, and an ability to operate a portion of those assets remotely creates a technological trinity that provides us a window into possible futures with respect to unmanned and autonomous weapon systems. This technological trinity is a powerfully seductive force for political leaders looking for ways to combat humanitarian crises, atrocities, and global extremism. As Georgetown University professor of law David Koplow describes it, when the cost of adverse consequences of action are reduced, it relaxes the "self-deterrence" usually present in decisions about the necessity of war or combat operations and increases the propensity for such actions.[22] As Martin Cook noted above, the perceived ability of precision and stealth to keep coalition forces out of harm's way lowered the bar for intervention in Kosovo. Yet Koplow warns about the overpromise of such systems: "Warfare is always dangerous and expensive, and it carries no guarantees."[23] Even the best systems will fall off target, and even our precision cannot "root out all opposition." In fact, we might find ourselves "rudely surprised by a rising local resistance" to our actions. He warns the seductiveness of air power might embolden "us to inject our forces into tasks that ultimately prove costly to finish by other means."[24] Or they might simply not work at all.

It would be hard to conclude the cruise missile strikes in Afghanistan in 1998 were effective in achieving the objective. Over the course of their thousand-mile journey, bin Laden simply went somewhere else. As we have seen previously, the use of air power in Kosovo, though eventually bringing Milošević to the table, did nothing to stop the ethnic cleansing that was the humanitarian rationale for the action in the first place. As Cook puts it, impunity gives leaders an opportunity for "employing force in circumstances that generate no friendly casualties, but which do nothing to redress the political causes that may be at the root of the conflict: a truly desultory and futile exercise of military power."[25] We will return to this problem later in discussions on impunity and the future of war. Consider, for now, the seductiveness of air power, its precision, and its current high degree of impunity. Can we draw parallels on the future of war?

With air power as an early example, how much lower will the barriers to war become if there is a capability to employ robotic or remotely controlled forces in future conflicts? Imagine the ultimate result of the public's call to do everything possible to minimize casualties and the presidential or congressional reaction to it. Author and professor of international relations Andrew Bacevich and journalist and author Robert Kaplan warn, "Ultimately a doctrine that relies on the antiseptic methods of warfare may prove dangerously seductive. . . . [It] may tempt them to conclude that force need no longer remain the option of last resort, and induce them to employ their arsenal without due reflection."[26] Frida Berrigan, senior program associate at the New America Foundation's Arms and Security Initiative, says robots will enable policymakers to avoid a draft, high casualty figures, and political debate about the proper use of military force.[27] Journalist Eric Stoner concludes in the *Boise Weekly*, "In effect, by reducing the political capital at stake, robots will make it far easier for governments to start wars in the first place."[28] And for an even darker representation, journalist Steve Featherstone writes in *Harper's Magazine*: "Within our lifetime, robots will give us the ability to wage war without committing ourselves to the human cost of actually fighting a war. War will become a routine, a program. The great nineteenth-century military theorist Carl von Clausewitz understood that although war may have rational goals, the conduct of war is fundamentally irrational and unpredictable. Absent fear, war cannot be called war. A better name for it would be target practice."[29] And this leads to what is probably the most disturbing part of the dimension of impunity and politics: whether leaders and presidents view remote killing as military force or war at all.

Mary Ellen O'Connell is the Robert and Marion Short Professor of Law and Research Professor of International Dispute Resolution at the University of Notre Dame. She has been outspoken on the use of drone strikes in targeted killings outside of "defined" war zones, such as in Pakistan and Yemen. I interviewed her in November 2010 fresh from a talk she gave at the USAF Air University in Montgomery, Alabama. E-mail traffic around the university prior to her visit teemed with speculation on what she would say and whether there ought to be a prepared response or a debate format. Academic freedom can sometimes clash with service politics.

Professor O'Connell told me she was asked about the psychology of remote killing but not in the way she expected. She thought the questions would go toward how the RPA operator views killing from afar but was surprised to note the questions were about the psychology of the president instead. How did he view such operations? Was it combat? Was it war? Was it military force at all or some form of extrajudicial enforcement? O'Connell admitted she had not really considered the possibility before but that the latter was a very intriguing question. She wondered why not go in on the ground or even fly fighter aircraft over Yemen and Pakistan? Is it because this would clearly be a military action in sovereign nations we are not at war with? Drones may simply be too easy, and they might be operated by nonmilitary personnel; if so, this places their use even further from a military act of force.[30] This is the likely concern of Special Rapporteur Alston in his study of targeted killings where he states, "The result of this mix has been a highly problematic blurring and expansion of the boundaries of the applicable legal frameworks—human rights law, the laws of war, and the law applicable to the use of inter-state force."[31] Other authors also worry about blurring the lines between military action and law enforcement.[32]

The "Global War on Terrorism," or more recently "Overseas Contingency Operations," has done its part to blur the lines between military operations and the law.[33] An earlier special rapporteur on extrajudicial, summary, and arbitrary killings called the case of the first U.S. drone strike in Yemen in 2003 "a clear case of extrajudicial killing."[34] Witness the more recent discussions about military tribunals or civilian courts for those accused of terrorist operations and the ongoing battle over the status of detainees at Guantánamo Bay. These discussions and the use of drones in such places as Yemen and Pakistan may be illustrative of how a president may view these operations. It is somewhat surprising that targeted killings in Pakistan have actually increased since 2008.[35] What does this say of the psychology of killing with impunity?

The ability to fly unmanned aircraft into the airspace of other states and the ambiguity about who is actually doing the flying and firing from several thousand miles away may allow one to think of this kind of operation in terms that are not specifically military. O'Connell's question is a good one. Why not go in on the ground or fly a manned fighter to do the job, or for that matter a special operations helicopter or fixed-wing gunship? Those actions are clearly recognized in the law of war as military uses of force. When no member of the armed forces occupies a cockpit and the "boots on the ground" providing the intelligence and target identification may or may not be uniformed military—either way they are conducting clandestine operations—is it plausible to conclude the operation is not a military operation at all? If it is not, is it conceivably not really an act of war? Those we are targeting may or may not necessarily be considered legal combatants and afforded the rights that status entails. It is possible our national leaders do not consider kinetic actions by unmanned vehicles to be firmly anchored in the realm of war. If this is the case and we continue along the trend lines toward greater autonomy in our weapon systems and increased use of unmanned vehicles, thereby granting ourselves greater impunity, we might not only be exacerbating the risk inversion. We could also be ceding enormous power to an executive branch finally free of the last constraint on proceeding to war—that is, the idea that it is even war at all. As Patrick Lin, director of the Ethics + Emerging Sciences Group at California Polytechnic State University, coolly states, "If it weren't for drones, it's likely we wouldn't be there at all. These robots are enabling us to do things we wouldn't otherwise do."[36] So what is a concerned public to do? We move now to the psychology of the citizenry.

The Psychology of the Citizenry

The last draftee entered the U.S. military in 1973 and registration ended in 1975. It was brought back in 1980, and not a single person has been called to active duty since.[37] The U.S. military is called the "all-volunteer force" and successfully sustained itself as such through minor altercations in Libya in the 1980s, Grenada, and Panama, and major operations in Kuwait, Iraq, the Balkans, and Afghanistan. Because the size of the military has shrunk, with minor perturbations, since the end of the Vietnam War and since there is no longer a draft or mandatory service, the military is less a reflection of society than it once was. Martin Cook says, "On the one hand, America's military is much more professional, better trained and equipped, and of higher quality than we have ever experienced. On the other hand . . . it is unrepresentative of the society it serves in many respects and, some fear, deeply alienated from that society."[38]

Very rough math shows the uniformed military personnel on active duty in the United States at any given time is approximately 0.5 percent of the total population. Roughly 26 million living Americans have been in the military at one time or another,[39] bringing the overall percentage of those in the general population with a military background to somewhere around 9 percent. Though we are flush with expressions of support and there appears to be a concerted effort to recognize the work our servicemen and servicewomen do, this too points to the gap between society and its military. We go out of our way to recognize that which we have no close personal knowledge of or investment in. I am appreciative of these expressions on one level, but on a deeper one I am concerned about what they may say about our connection to the society we serve. As one senior Marine officer said in speaking to an audience at the National Defense University, America's youth does not learn about military service from uncles, brothers, and fathers as it did when he was growing up.[40]

The public does not feel the burden of war like they used to either. Americans have not been asked to ration goods. They have not been asked to pay for war with higher taxes. In fact, they received a tax cut at the very outset of "the long war" and recently saw those measures extended. They have consumed iPads, Kindles, hybrid or fully electric cars, and iPhones—all innovations of the last decade—because factories were not required to concentrate on war matériel. As a friend of mine said during a discussion of his organization's place in U.S. war strategy, "We are a military at war, but we are not a nation at war." We have not been asked to share the sentiment Camus expressed about World War II when he wrote, "This is why, however vile this war may be, no one can stand aside from it. I myself first of all, naturally, since I can have no fear in risking my life by wagering on death. And then all the nameless and resigned who go off to this unpardonable slaughter—and who, I feel, are all with me as brothers."[41] If the society is alienated from its military today, what will the future bring?

Peter Singer believes it will grow even worse: "Unmanned systems (and their ability to carry out remote acts of force) erode the deterrent exerted by public sentiment, a decline already begun by the end of the U.S. military draft. . . ."[42] Michael Ignatieff is concerned about the loss of checks and balances in what he calls "virtual war,"[43] a term he defines as "war fought in the search for moral impunity . . . war that attempts to be prosecuted without risk to your own side."[44] These are wars that no longer involve national survival, no longer involve the citizenry (due to lack of conscription), no longer require democratic consent (due to bypassing of representative institutions), and no longer require the entire economic system.[45] That is a pretty good description of the world we occupy now, and there is no reason to believe the advent of autonomous weapons or increased use of unmanned systems will cause the general public to become personally reinvested in questions about the use of military force.

It is the duty of a citizenry to communicate its will to its elected officials in the run-up to war. It is the duty of a free press to report on the rationale and public opinion. Citizens elect their representatives in the legislature and their president in the executive branch, but their duties are not complete upon casting a ballot. Democracies only work when the will of the people is expressed and known. Personal investment in the risky area of warfare is one way the electorate has managed to stay engaged in the political process. The electorate should restrain the frivolous use of military force. Until now, democracies have required basic agreement, or at least a lack of disqualifying disagreement, before going to war.[46] That may all be changing.

Democracies have short memories. In the United States we take that concept to a whole new level. The farther we are from the beginning of something—anything—the less we remember and the less we care. Recall the attempted so-called Christmas Bombing of 2009 when a man attempted to bring down a plane by setting his explosive underwear on fire. Remember the calls to rush imaging technology into the airports? Then, a mere one year later, came the now infamous YouTube video of a passenger expressing his displeasure at the placement of anonymous Transportation Safety Administration hands after he refused to submit to the electronic scan of that same imaging technology. Whether attempted terrorist attacks or "long wars," our interest wanes with the passing of time, but this is hardly a new phenomenon. In 1891 Henry Sidgwick warned, "An important incidental evil of a widely-extended war is, that the restraining force of public opinion on the belligerents is inevitably much reduced by it."[47] Combine a natural apathy with technologies that take the sting of war away, and we are left with a dangerous potion leading to secession from the necessary duties of the electorate.

In *Ethics and War in the 21st Century*, Christopher Coker warns that our system of creating technology has so far allowed a dramatic increase in the productivity of the single soldier and has therefore demanded less of the individual operator. Historically this has coincided with a corresponding increase in what has been demanded of a society: "The problem with robotics is that it demands even less of the individual soldiers it is threatening to replace. In increasing the distance between ourselves and our enemies, robotics is also demanding less and less of the societies that send them out to battle."[48] We have moved far beyond Camus's sense of shared brotherhood. Americans are a coddled lot, living in far and away the most comfortable society the world has known and ever farther from those hard choices of a hard life that brought us here. Nothing is really asked of us in fighting two major combat operations even with the force we have today. How could it be any different if our soldiers were robots or our carbon-based ones were kept far from harm?

Virtual war, or autonomous or robotic war, alters our collective reasoning about its cost. Without feeling it at all, Americans have funded the enormous explosion in robotic technology cataloged in previous chapters and allowed it to be deployed and take part in operations that, just a generation ago, might have been considered against basic American values and ideas about how to conduct a war. The irony is we have done so all the while believing we know more today about the essence of war than at any time before.

The images of war come to living rooms every night and have been doing so, at least while the nation has been engaged in combat operations, since the Vietnam War. After a period of my father being away during that war, my mother forbade me to watch the evening news. I was five and had been having nightmares of seeing him get killed. But what I saw on the screen then, what I watched during the Gulf War, and what I've seen in every conflict since is not really war. Coker says what we see is not, as people think, exhaustive; it is only a glimpse, and it limits the imagination by making things visible. Yet sometimes what we see compels us to act without considering the enormity of what we begin. He explains that while "televised images of human suffering may impel people to act, they do not provide any insight into anything that approaches irony, conflict, or dilemma—the dilemma, for example, of having to act cruelly in the name of humanity or humanitarianism or at the very least accept the need for a human cost. That is the stuff of the tragic sense of history." In the end we see war as that five-year-old of decades ago: "We experience war without understanding it. Our subjective understanding of it is not what it was."[49]

The technology driving impunity in warfare creates an intoxicating elixir—an opiate for the masses, if you will—promising immaculate and virtual war. It claims wars can be fought with minimal risk, minimal or no casualties, and no burden on the populace at large, all while being broadcast with the clarity we have come to expect from our high-speed-Internet-connected computers and high-definition televisions. We can see suffering without experiencing it and decide to act without paying for it. We experience war without having to understand its ferocious and utterly unforgiving nature. And because we can do all these things, we can also abdicate our responsibilities as a citizenry to act as a restraining force on an increasingly unconstrained political leadership. In doing so, we make conflict ever more prevalent. As Eric Stoner explains in his piece "Attack of the Killer Robots," "By distancing soldiers from the horrors of war and making it easier for politicians to resort to military force, armed robots will likely give birth to a far more dangerous world."[50] Former Army chief of staff Gen. George W. Casey Jr. claims we are in a state of "persistent conflict," whereby regional friction continues to fester into armed interventions, but if it is so, it is likely of our own doing. The recent adventurism in kinetic foreign policy over the last decade may be symptomatic of an insidious retreat from restraint in war; we have implicitly chosen persistent conflict by our own action and inaction. Sidgwick's warning is both self-fulfilling and self-propelling as we become accustomed to constant war. Unfortunately, "the coming robot army" will do nothing to change it.

Effects on Civil-Military Relations

Imagine if you can—and I know this is difficult—that trouble is brewing in some part of the world where stability is very important to the vital national interests of the United States. Imagine too that, for reasons that could take hundreds of pages to explain, the people in that part of the world are unlikely to view U.S. intervention in the same light as the French did when American general George S. Patton was ordered to allow Free French leader Charles de Gaulle to march into Paris first. Due to a history of these kinds of problems over the past thirty or so years, the defense establishment got pretty good at thinking up new ideas, building and testing things on what really amounts to a shoestring budget, and then putting them on a shelf—in places with very cool names like DARPA, the Air Force Research Laboratory, Army Research Laboratory, and the Office of Naval Research (ONR)—and waiting for their time to come. We have research organizations—the DOD's "venture capitalists" as one DARPA technician described them—whose job it is to create solutions looking for problems.[51]

The concept of DARPA engaging in venture capitalism is a pretty good analogy. The ideas venture capitalists are looking for are those things no consumer ever thought he needed. In fact, he had never even conceived of a place for such a product in his life until, all of a sudden, it was the latest and greatest thing everyone had to have. Think of all those gizmos that start with a little *i* in pockets, backpacks, briefcases, and future-museum-like stores all over the globe. These are transformative, possibly disruptive technologies that change the way people listen to music, gather information, and even interact with each other. These are part of business strategies that seek to change the rules of the game, and for those who are successful there is enormous profit to be made. Venture capitalists look to create markets where there were none before. They are, in "Pentagonspeak," in the business of "generating requirements."

Early in my career I typed papers, and my boss actually signed them—with a pen. I then ran them around the base for other people to sign. Now we couldn't do our jobs without e-mail and the computers we use to send and receive it. Business and much of government runs on smartphones. These "requirements" were once just ideas in visionaries' minds and were brought to life with the backing of a few wise venture capitalists.

Versions of PackBot from iRobot (a different small *i* company) began its life as a DARPA project.[52] So did General Atomics' Predator drone.[53] The Predator is now part of what is known as a "program of record," meaning it has a funding placeholder in the budget every year (it still must be defended each budget cycle) for operations, maintenance, and life-cycle support. Programs of record are the Holy Grail for defense contractors, as they represent a relatively safe funding stream. They are the lottery win for once-small companies like General Atomics and iRobot, whose products began as "advanced concept technology demonstrations." These two programs are the faces of unmanned technology in the fight today in Central Command's area of responsibility. A DARPA technician recently expressed a sense of dismay over the USAF's early apathy about the agency's Predator predecessors.[54] Hindsight is sweet luxury. They are now widely viewed as very successful first steps into unmanned military operations. Arguing with success is rarely meaningful, and it is surely inefficient. It is also not the intent here. At issue is something far more insidious, maybe even ominous, about the combination of technology trends and the creation of solutions in defense technology development for problems we have not yet encountered.

I began this chapter with a story of the very fine line professional military officers must walk in discussions with their civilian leaders as they provide military advice. Our system is based on civilian authorities making decisions and military officers executing those decisions, but as I have gained more experience, I have learned to accept that not all things are so black and white. I turn again to Martin Cook, who argues for a dual nature of military advice—both the "how," which we are much more accustomed to, and the "should."[55] This is that fine line very senior officers walk. Elected civilian leaders must decide, but their decisions must be informed by the practicality of military options that only military men and women can provide. Despite the Hollywood caricature of warmonger generals, they are often the ones who advise against military options.

An excellent example is former chairman of the Joint Chiefs of Staff Gen. Colin Powell commenting in the beginning of President Bill Clinton's presidency about dealing with the problems of Bosnia. He repeatedly said we *should not* commit military forces until there were clear political objectives. Secretary of State Madeleine Albright then exclaimed, "What's the point of having this superb military that you're always talking about if we can't use it?"[56] Few would question the appropriateness of General Powell's advice here, except perhaps Secretary Albright, but her comment is also instructive. It is a warning journalist Stephen Wrage spells out in "When War Isn't Hell: A Cautionary Tale": "If there are unusually useable weapons in the arsenal, there will be unusual pressures to use them."[57]

The drive for impunity in warfare has granted military officers the ability to provide to the legitimate authority "unusually useable" options for military force—or whatever force they believe it to be. In doing so we have crossed, or at least are slowly extending a toe across, a line between *ad bellum* determinations and *in bello* uses of our means of war. Consider the following from Professor Harald Müller discussing the so-called revolution brought on by precision ordnance:

> The RMA [Revolution in Military Affairs] and related changes in strategy and tactics can be read as a deliberate effort to draw the lessons from the Vietnam disaster where a vast power asymmetry did not translate into the expected victory. Looked at in this way, the "Revolution in Military Affairs" almost amounts to a conscious attempt by the military leadership and their political masters to evade the inhibitions that cost-risk-benefit calculation by an enlightened citizenry in a democracy impose on political decision-making (Mueller, 1989; Luttwak, 1996). The RMA—if all dreams come true—is tackling all the weak spots in the determination of a democratic population to agree to going to war—if

otherwise good reasons are given why to do so (Müller and Schornig, 2001a, b).[58]

Now I do not agree that this blurring of the line or evasion of inhibitions is a conscious effort by senior military and civilian leaders, but whether it is or not is completely irrelevant. What matters is whether the drive for impunity changes this calculation and, consciously or not, blurs the lines between responsible parties to *ad bellum* and *in bello* considerations. I believe it does, and I offer this previously noted example to highlight impunity's power and asymmetry's supporting role in the decision process.

In the spring of 1945, President Harry S. Truman must have wrestled with how to deal the final blow to Japan. There seemed two courses of action: an invasion or the low-probability-of-success nuclear weapon that great minds were working on in the middle of the New Mexico desert. The first would likely bring a high cost in American and Japanese casualties, the latter a high cost only in Japanese casualties. On the first day of the Potsdam Conference, the Manhattan Project came to success, and Truman now had an option that did not involve the inevitable deaths of Americans. He was reasonably sure the Japanese did not have such a capability, therefore reciprocity was a very remote probability. The rest is history.

This is not a completely pristine analogy, I realize. The problem set for Truman was how to make the Japanese capitulate, and all options at that point were military. There was no question of "Should we try to win the war with Japan?" But for a moment consider the essence of the example. On the one hand was the brute-force, boots-on-the-ground method. On the other, delivered by the greatest technology demonstration in history, was a brutal but scientifically exquisite solution offering complete impunity. You choose.

There is a danger in upsetting the civil-military dynamic in pursuing this kind of immaculate war, and it feeds both the psychology of our leaders and the electorate. Impunity tends to make the "should" questions of *ad bellum* determinations closer to the "how" questions of *in bello* actions, and it is exacerbated when an overwhelming asymmetry exists. These effects are insidious. Our leaders are likely influenced on a subconscious level. The effects do not require some secret desire of military officers or civilian leaders to lower the inhibitions for war—I am convinced this does not exist. It does not even matter if casualty aversion, acting as a brake on decisions to use force, is even real. As Coker says, whether or not the zero tolerance of casualties is accurate, the perception is real, and it shapes military planning and "long-term force development."[59] We continue to pursue impunity because we care about the lives of our soldiers and their families.

As our military options continue toward lower and lower human costs, there is a danger of having "unusually useable" weapons. It may drive a perception that we can act whenever and wherever we choose[60] without due regard for the risks. It may impact what Koplow calls our "judicious self-restraint."[61] How impunity affects our military advice to our civilian leaders is something we ought to consider with some depth. We have to come to grips with whether our desire to keep our soldiers out of harm's way is worth a possible erosion in distinction of military advice and whether our procurement of advanced technologies alters the dynamic between the citizens they defend and the authorities they serve.

No Man's Land

Impunity seems to have discernible effects on the politics of war. It affects the psychology of leaders and the electorate, and strains our concept of appropriate military advice to our civilian leaders. The drive toward lethal unmanned warfare can only magnify these effects as, eventually, the human cost of war may be assumed away in favor of an autonomous army of robotic systems. What would John Ruskin, whose quote opened this chapter, have had to say about the "goodliness" in a game of war played with a multitude of nonhuman pawns instead? He probably could not have conceived how this may be or that it would affect the mechanics of war in any discreet way, but he would certainly understand how it might change the political dynamic that inevitably surrounds decisions about war.

Inasmuch as the drive for impunity in war reduces the risk to combatants and noncombatants of technologically superior states prior to the risk inversion, it lowers the barriers to uses of force. Impunity tends to subvert the last-resort requirement for acts of war, since its primary constraints are the potential for casualties and the need to seek the combatants' consent before resorting to armed conflict. Asymmetry, precision, and remote warfare form a technological trinity of willing enablers of impunity. The threat of reciprocity also hangs over the heads of those whose job it is to decide for or against war, and such decisions tend to be easier when there is a low probability of retribution. Precision allows leaders to believe dirty work can be done with surgical cleanliness and efficiency, though it is not always the case. Remote warfare enhances the dream of near-total impunity, but it also adds ambiguity to the legitimate authority's views that such remote strikes are even considered warfare at all.

The most troubling aspect of the psychology of the leaders is whether they even consider the use of lethal unmanned operations as acts of military force. Because of the clandestine nature of the operations on the ground and the fact that no uniformed personnel occupy the UAV, it may be easier for leaders to consider such uses of force as something other than military. It is unclear whether they think of such operations in terms of the acts of war many others consider them to be. If they do not, there are wide-ranging ramifications for those conducting the strikes and those who are being targeted. Are either really combatants? If not, what is their status? Such a view conceivably stretches and blurs traditional legal frameworks to deal with acts of force. With the lagging nature of the law discussed previously, we might be in a no-man's-land between law and war with no good way to determine how we will one day be judged. Perhaps we should hope for a public as apathetic as it seems to be in acting as a restraint to our use of force.

The U.S. citizenry and military are increasingly less reflective of each other in a time when the public is no longer required to feel the weight of responsibility for its nation's forays into war. Because the public has not been required to feel the full cost of our recent wars, it is less inclined to engage in the civic duty of reasonable discourse on uses of force. When the executive can decide to use force of its own accord, possibly characterizing it as something other than war, the public is farther removed from the decision process. Add impunity to the mix where there is a decreasing chance of casualties, and the public virtually secedes from its responsibilities. A disengaged and disconnected electorate fuels the most ominous prospect in the dimension of impunity and the politics of war.

Senior military officers must walk a fine line between advising on the practicality of uses of force and wading into questions on whether such force should be used. It is not a black-and-white situation, and in the end they must do some of both. However, the lines of authority are clear. Only the legitimate civilian leaders can decide to go to war or use military force. Unfortunately the drive for impunity has also blurred this line, creating "unusually useable" weapons and leading some senior civilians to wonder what good a fine military is if it cannot be used. Our desire for the latest in technology, to the point where we purposefully fund agencies that dream up solutions for problems we have not yet faced, threatens to erase the distinction between accountability for *ad bellum* decisions and *in bello* uses of force. The advent of robotic warfare will only make this worse. We must eventually deal with the tough issues surrounding our desire to keep our soldiers out of harm's way and the possibility of this desire destroying the distinction of appropriate use of military advice.

The psychology of the leaders and citizens, as well as the impact to appropriate civil-military relations, are the complex issues involved in impunity and the politics of war. Procurement of systems that tend to increase the impunity of remote warfare, asymmetry, and precision—the technological trinity—all at the same time are likely only to magnify the problems highlighted above. It may be, as Ruskin warned, there is simply no virtue or "goodliness" when the game of war is played with a multitude of small pawns of any kind.

6

Impunity and the Warrior

And what I understand most clearly is the contrast which you make between your own readiness to die and your revulsion at the idea of other people's death. This proves a man's quality.

—Albert Camus, letter to a man in despair[1]

What happens to every major civilization? At some point they civilize themselves right out of warriors.

—M.Sgt. Mike Gomez, USMC, UAV pilot[2]

On the first day of my second most recent foray into academia, at the Air Command and Staff College in Montgomery, Alabama, the commander of the student squadron stood on the stage to address his new charges. It was his job to inspire us for the coming year of book learning while many of us felt, after just shy of three years at war, that there was still more for us to be doing "out there" in trying to win the thing. I've forgotten most of what he said, but one thing stuck to some part of my gray matter in a way that will probably defy my coming forgetfulness. He said something like, "Good morning, warriors! No, I don't really like that term—you're all combatants." I found the statement odd, but I'm wondering if he was on to something. Of course not all of us were combatants either. It is a legal term, and I'm sure the chaplains in the class were more taken aback than I. Still, as I look back it seems fairly certain not all of us with combatant status could be fairly judged as warriors either.

Warriors, Combatants, or Something Else?

There are a host of career fields in our nation's military, and most of them are in support functions for our combat arms. Many have analogous career paths in the civilian world, and in none of them is killing job one. Even among the military fields where it is the primary job, not all who occupy the field would likely consider themselves warriors. There are those who are likely born to their chosen profession and who see what they do as a life-affirming calling, and there are those who view it simply as a job. Among those who shared my particular niche, there was a saying: "There are fighter pilots, and there are pilots who fly fighters." I suspect there are similar sayings in other fields. There are warriors, and there are combatants who pull triggers. In our technological way of war, there is certainly valid debate over whether any of us qualify as warriors. There were indeed more combatants in the room that day than warriors.

You might be thinking, "Fine—so what?" It's an excellent question. Are we beyond the need for a warrior ethos in the early twenty-first century? I will argue that we are not, but first it is important to understand what I mean by the term. "Warrior ethos" is a difficult concept to define, perhaps even harder to grasp. It begins with the manner in which those in the profession of arms view their chosen craft. First it must be considered a craft; more accurately, it ought to be considered an art. It is one that takes many years to cultivate, much like an apprentice takes years to learn a chosen skill. A warrior is technically proficient but views war more philosophically. A warrior is more artisan than technocrat. Perhaps this is why we find ourselves asking whether a warrior culture survives and is relevant today. These views tend to collide. A warrior also feels called to his or her chosen path and sees the art of war as an existential act. Christopher Coker explains that this existential nature of war is "the reason that warriors are prepared to die (as opposed to kill)."[3] He notes there is a sliver of this culture attempting to hang on in the U.S. military today and explains the warrior ethos as that which "still celebrates combat as the supreme expression of military life."[4] It goes beyond duty, but it is problematic as we move into an age of robotic and unmanned warfare. It forces us to answer the question "What is combat?" This is not as easy as it first seems.

The rise of unmanned, autonomous, and robotic warfare forces the services to answer tough questions and challenges service culture. The extent to which robotics affects whatever warrior culture is left in twenty-first century Western militaries is as yet unknown. Killing with impunity almost certainly has a psychology all its own, and we are just now beginning to investigate it. We will attempt to explore whether killing by video screen detaches the operator from the act, as Dave Grossman, the author of *On Killing: The Psychological Cost of Learning to Kill in War and Society*, claims, or "enhances" the experience, as a few RPA operators say it does.

There is also an issue of living in two worlds nearly simultaneously—the world of war and the world at home. When I deployed for combat, I left my family and all the minutiae, just living a normal life entails. I was focused on the tasks at hand and was left with my own thoughts in the quiet solitude of small quarters or with the superb camaraderie of shared experience during after-duty hours in an eight-man tent. Remote warriors live in "combat" for twelve hours a day, then go home to household chores, homework, sometimes-sick kids, and some sense of normalcy.[5] But how normal can it be? It is fair to say we just do not yet know the full effects of distant killing and the "two-lives paradigm" it produces. Additionally, due to the risk inversion, remote combatants may be putting their families at far greater risk. We have not considered how to handle "combat" deaths in such places as Nevada's Creech Air Force Base, from which drones are operated, or even in nearby downtown Las Vegas.

And then there are practical matters affecting the warrior or combatant. In order for killing to be legal in war, it must be done by legal combatants. In the case of the Western way of war, and codified in the law of war, this means only uniformed military personnel are allowed to take lives. In other words there must be military control over lethal force, but autonomous warfare severely blurs the line of appropriate military authority. There are concerns about the deskilling of the military as more combat functions are handled by machines. Our ability to build and maintain true experience in the art of war may face serious barriers as we consider the degree to which we allow autonomous weapons and even unmanned vehicles to take part in lethal actions. War has, up to now, been a chaotic endeavor with a severe bent toward the irrational. It has so far defied scientific characterization and explanation. Are we conceivably on a path to "lab coat warriors," uniformed programmers who fill the square for military control and send their "bots" on their merry way to war, all while having no practical experience in the ebb and flow of mortal combat? Is it possible to maintain the intent of this requirement in a world where we do not consider experience in a physical battlespace as a prerequisite to order our forces, whatever form they may take, into the breach? Even if this extreme never comes to pass, there will be problems growing the right kinds of generals with the right kind of experience as robots become more numerous and humans on the battlefield become less prevalent.

Again, you may be thinking, "So what?" The intent here is not to say these issues are permanent roadblocks. It might be possible to solve the practical matters. We might have to learn to live with the psychological effects of remote killing, including second-order effects. It remains to be seen whether maintaining a warrior ethos is even relevant anymore, though I hope it is. Unfortunately, I will not be able to present solutions to all of these issues. In some cases there are probably not enough data to draw valid conclusions. In others, I simply lack the expertise. One thing is certain, though: those who do have expertise and can gather the appropriate data cannot solve the issues on any reasonable timeline before they are overcome by too-early fielding of these technologies, unless we confront them now with clarity and transparency. With that modest hope and a great deal of trepidation, I turn to impunity and the warrior ethos.

The Existential Nature of War

In beginning this project, I recognized I was going to be at a distinct disadvantage, depending on the tack the research took, based on the wings on my chest and the duty history stacked up over the past twenty some years in my personnel records. After all, a fighter pilot expressing misgivings over unmanned vehicles in general and unmanned aircraft in particular is clearly living in an overglorified past and fearing for the future of what Peter Singer describes as the Air Force's leadership DNA.[6] Spending a year at the DOD's only senior service school committed to teaching the intricacies of resourcing the National Security Strategy meant taking more than a few spears because of such acquisition debacles as the next-generation tanker, incredibly long procurement cycles for projects such as the F-22, and growing dissent over the validity of the most expensive defense acquisition program in history, the F-35 Joint Strike Fighter. These issues give the casual observer cause to believe the Air Force somehow missed the memo on the future of war and still believes it is preparing to fight a long-dead enemy.

While the uncertain possibility of the future of war clearly *can* be overstated, there is no denying such concerns play on the minds of current and future Air Force leaders. It makes for fertile ground for those who see the perceived reluctance of services to press forward with unmanned systems as simply a culture problem[7] belonging to what the DOD's *Unmanned Systems Roadmap* describes as those "pockets of resistance" that must be "eliminated."[8] Even defense policy analyst Andrew Krepinevich jokes that "no fighter pilot is ever going to pick up a girl at a bar by saying he flies a U.A.V. . . . Fighter pilots don't want to be replaced."[9] In the end these are too-easy retreats to simplistic arguments about matters that deserve all the intellectual capacity we can muster. Allow me to suggest there might just be something much deeper at play.

Unmanned systems capable of lethal action subvert what it means to engage in combat and confront our sense of what it means to be a warrior. This is a culture issue for sure, but on a whole different plane than the "I'm a fighter pilot—how do you like me so far?" level. It is very important to understand that experiencing combat and being a warrior are two very different things, and our misunderstanding of their relationship causes friction in the ranks and in our own sense of who we are as war fighters. As Singer says, "this disconnection from the battlefield also leads to a demographic change in who does what in war and the issues it provokes about a soldier's identity . . . or status . . . or the nature of combat stress and fatigue."[10]

Consider these two views. In recent discussions about targeted killings and our ability to strike from afar and with total impunity, a senior officer and former fighter wing commander remarked, "Where's the chivalry in that?" Then there is a young officer just out of the Air Force Academy who speaks with wonderment about how flying Predators is seen "as this geeky thing to do" despite the fact that its pilots have seen far more combat than fighter pilots in recent years.[11] Misgivings about unmanned warfare are not about pickup lines or shiny stars lined up on epaulettes. They are about a nearly dormant and continually repressed sense of our warrior spirit.

Author George Leonard, writing in the introduction of Richard Strozzi-Heckler's *In Search of the Warrior Spirit*—an amazing book about this aikido master's experiment in training U.S. Army Special Forces soldiers in the ancient ways of the martial arts—says, "I can't imagine anyone from here on out writing a book about the warrior in a technological age, in a nuclear age, in a terrorist age, without referring to Strozzi-Heckler on the subject."[12] I cannot imagine it either, and as this work is ultimately about the warrior in a technological age, I will take him at his word. Strozzi-Heckler describes various warrior traditions and says, "These historical and mythical warriors found their strength and integrity by defeating their own inner demons, living in harmony with nature, and serving their fellow man."[13] As Coker explains, citing the ancient Greeks, they "offered warriors the realization of their own humanity. They found in war a master text by which they came to know themselves better."[14] It is a far more personal concept of war than we allow for today.

What is "the warrior spirit?" Allow a perspective from one untrained in the ways of Strozzi-Heckler and Leonard's Aikido or in the social sciences, but nonetheless from one who has attempted to find such a spirit for over two decades in the midst of the technologies we discuss. The warrior spirit is a sense that what a warrior does in war and how he comes at it on a personal level transcends the cold rationality of performing a mission, completing an objective, taking a hill. It neither ignores nor celebrates the necessity of taking human life. It understands sacrifice only on a personal level and in relation to fellow warriors, and therefore does not expect or desire any recognition of status other than that of "warrior." It sees combat as the ultimate and artistic expression of a life spent in its preparation. Combat for the warrior is an intricate dance, a test of personal will and technical skill, played for the highest of stakes where form in means is as important—perhaps more so—than the desired end. The end state must be achieved, to be sure. That is the reason for military action. But those who can perform it with more finesse and elegance are better respected for their mastery of the craft.

Leonard describes this idea in the context of Strozzi-Heckler's black-belt test: "It was like one of those sporting events that are later memorialized, perhaps a World Series game or a bullfight, during which every last spectator realizes at some level that what is happening out on the field is more than a game, but rather something achingly beautiful and inevitable, an enactment in space and time of how the universe works, how things are."[15] It is this personal and aesthetic quality of war we risk losing with unmanned or robotic warfare. Would we recognize a no-hitter pitched by a pitching machine with the same awe as we do when we see the battle between the man on the mound and the men at the plate? Would we feel the same sense of loss when in the ninth inning the last at bat slings one into the upper deck in center field? Our sense of the warrior and his sense of his place in war are not trivial matters, for it is the aesthetic quality of war that helps ground it as a human activity.

Tragedy, in the ancient sense of the hero's fall from grace and eventual understanding of some deep philosophical lesson, has long been a part of the world of art. It is part of the human experience, maybe even integrally so. The degree to which automated warfare removes us from this sense of the tragic alters our ability to understand the essence of what we do in war. Coker describes it this way: "Without a sense of the tragic, it is difficult to maintain a humanistic understanding of war or, for that matter, for soldiers to see themselves as warriors."[16] The warrior spirit is intricately linked to this sense of the tragic and drives war into the personal realm. The warrior strives to maintain the nobility of the tragic hero before the fall. Strozzi-Heckler says, "It occurs to me that part of being human is the longing, or perhaps even need, for the experiences of courage, selflessness, heroism, service, and transcendence."[17] It might be this need, the recognition that warfare with impunity directly affects it, and the manner in which we have attempted to address it that cause the friction Singer and I have described above. How do we know who is at war and who is not? How do we address the dichotomy between the senior officer's view of chivalry and the junior officer's view of combat described above?

I have already described the morality and legality of the requirement for risk to be considered a combatant, and I will not dredge it up here, but consider this view from an officer in special operations recently returned from combat duty in Iraq who, when asked if he thought a Predator operator was at war, said, "No. He doesn't meet my definition."[18] Yet the Air Force, the furthest along the path of unmanned warfare today, tends to bend over backwards attempting to contradict this officer and, to be fair, the thoughts of many aviators of the Air Forces's combat air forces.[19] Consider these comments from the wing commander at Creech Air Force Base: "It's not really 8,000 miles away, it's 18 inches away. We're closer . . . than we've ever been as a service. There's no detachment. Those employing the system are very involved at a personal level in combat. You hear the AK-47 going off, the intensity of the voice on the radio calling for help. You're looking at him, 18 inches away from him."[20]

But there is a problem with this view of combat. On a grand scale, if this is combat, so is sitting in the Pentagon or any number of headquarters around the world, with the ability to see the live video feed. Perhaps, for that matter, so is sitting in an armchair and watching that same video on the evening news. In our struggle to define what it means to be in combat in our postrobotic world, we run the risk of diluting the concept beyond any recognition or meaning.

During World War II, the Japanese attempted to turn their nation, by conversion of each individual citizen, into a warrior nation by harking back to the samurai tradition. In doing so it lost a little of that great tradition and ultimately failed to achieve the hopes of its military leaders. As Coker explains, "A whole nation cannot be made into warriors; in its attempt to do so, the high command subverted the very notion of the warrior's honor."[21] There is a danger of doing the same thing in attempting to stretch the concept of what it means to take part in combat or worse, misconstruing what it means to be a warrior if we continue to proliferate armed unmanned systems throughout the services. An "everyone a warrior" mentality fails to respect an admittedly fading but still proud warrior spirit that remains at the core of our services' combat arms. There must remain some kind of personal sense of the tragic in a high-stakes contest in order to maintain a meaningful view of warfare. For those involved in it, a personal investment, an understanding of the potential for personal growth at war, insulates us from the possibility of a too-easy acceptance of killing. We ought to be fearful of those who would kill simply out of a sense of duty without considering a deeper meaning. It is probably better for the one with the finger on the trigger to also be someone who takes war personally. This requires us to double back once more. In order to understand the existential nature or war, at some point we must return to the concept of personal risk.

As I have described, Michael Walzer says no principle of just war theory bars distant warfare (he is not speaking in terms of impunity here but pure distance),[22] but Camus argues one should not be allowed to kill unless one is prepared to die. This certainly does not square with our views about keeping our soldiers far from harm, says Walzer, "Yet there is a wider sense in which Camus is right."[23] How can the desired needs for courage, selflessness, service, and transcendence that Strozzi-Heckler mentions be met in a world of impunity? Coker says the Western way of war is nearly completely instrumental and based on what it takes to kill members of the other side, whereas "non-western strategies ask a very different question: What does it take to persuade soldiers to die for their beliefs?"[24] It is in this sense that Camus is right, that for the warrior there can be no sense of accomplishment, no growth, without the test of mortal danger. Without it what he experiences is not war but simply a game.

When confronted by a Pentagon staffer about this idea, incredulous that it was coming from a pilot whose ilk had shown so little acceptance of risk over Kosovo and amazed that I would consider it a requirement to die to take the step beyond combatant and be considered as part of a warrior class, I had to note a subtle point. One only has to be willing to die; one does not have to complete that particular task. In the intricate dance of air-to-air combat, most pilots would like nothing better than to kill their adversaries from beyond visual range, but it does not always happen for reasons that are a microcosm of the study of Clausewitz. There are multiple layers of tactical plans that, if needed, can lead to a one-on-one, high-g[25] dogfight—what we refer to as "a knife fight in a phone booth"—in air combat's equivalent of hand-to-hand fighting. For those air forces advantaged in long-range weaponry (as U.S. air forces are), each closing mile at high subsonic or supersonic speeds means increased risk and higher probabilities for failure. Similarly, when aviators primarily dropped free-fall weapons from relatively low altitude, there was a saying about the critical five seconds "on final" during the delivery phase when the wings were level, the aircraft was in a stable dive attempting to smooth out the weapon solution, and the enemy was very capable of calculating an easy solution of his own. It was said those five seconds were "what you were paid for." It is no different today, and some precision-ordnance delivery options require far more time in relatively predictable flight.

What these examples share is a sense that we must be willing to do what is necessary, even at great personal risk, in order to engage in the personal test of will and skill at the heart of the artful view of battle. They show there still lives a core of those with the warrior spirit, a cadre capable of asking what it takes to persuade men and women to be prepared to die rather than simply kill. It is this concept that allows pilots to fly into known surface-to-air missile envelopes in order to deliver the proper ordnance for the job—weapon effectiveness has an aesthetic quality all its own—or rescue friendly forces in trouble. It is the reason men will burst through doors in insurgent-held buildings or throw themselves on live grenades or injured comrades during a firefight. It is a personal sense of the tragic and a view of war as "achingly beautiful and inevitable" that sings to its existential nature. It is easy to recognize those who grasp its depth. It is equally desirable, knowing the inevitability of war, to follow those who find personal meaning in it. To do so holds the promise of avoiding wanton killing and the ends-focused mind-set that bore the destructiveness of attrition warfare.

I believe the warrior ethos remains an important concept in modern warfare, but there is no guarantee that it will, or even needs to, remain so. The moral and legalistic arguments made in previous chapters notwithstanding, we may very well be moving into era when what I have described as the concept of combat is simply a quaint remembrance of the past. Perhaps it is already so. Unmanned and robotic warfare might accelerate the demise of the warrior spirit, or it might force a new understanding of this ancient concept. We would do well to think about it deeply, perhaps even meditatively, before we find ourselves overrun by the consequences of our action and inaction. Simply redefining combat as everything we do and see in war, or telling everyone in uniform they are warriors, is not the answer.

Xenophon's warning from the fifth century rings true: "There is no beauty when something is forced or misunderstood."[26]

"No Community of Fate"

The warrior ethos is only one part of the warrior dimension of impunity. In this section we move away from the philosophy of the warrior and get progressively more concrete by discussing the effects of virtual war on the warrior both in terms of what he sees in the conduct of operations and in how he deals with the aftermath. Then we will examine the ramifications of the risk inversion hypothesis and how it may affect the mind-set of the remote warrior. Ultimately this section will end in the more tangible world of the practical implications of impunity and the warrior, but first there is more to explore in the mind of the warrior.

Ever since the Gulf War, the term "CNN effect" has been in the collective vocabulary. It is the ability of a wide public to see, through the lens of the cable news outlets, their wars being waged with a simultaneity they had not known before. Of course, now "instant news" is broadcast via the Internet and captured by millions of handheld devices. Our own technological trends in warfare allow us to play back audio and video from weapons impacting targets, from young Marines facing insurgents, and from cockpit communications as helicopter pilots decide to fire. These images make us feel like we are part of the action in one sense yet may detach us from the consequences on the other. It would be fair to say the jury is still out on what the CNN effect does to our reaction to warfare. First, let us consider the case that it causes detachment.

There is, in some of the literature on distant killing and killing with impunity, a conspiracy-theorist bent about the technological trends in warfare. This slant is apparent from Harald Müller's preceding statement that the Revolution in Military Affairs was a conscious effort to lower the barriers to war. Consider Grossman once again, who says, "There is within most men an intense resistance to killing their fellow man. A resistance so strong that, in many circumstances, soldiers on the battlefield will die before they can overcome it."[27] There is an argument that says the way the military has solved this quandary is to "introduce distance into the equation."[28] There is probably not anything quite so calculated involved in this trend of technology, but there is still a reasonable perception that distance decreases the resistance to killing. I have made the case above that impunity in war is likely to lower the bar for using force and may even alter our notion of whether we are at war at all. Is it as likely to remove a sense of consequence from those pushing the pickle buttons or pulling triggers?

During the Gulf War we watched bombs falling on targets all around Iraq. We saw the antiaircraft gunfire. We could hear the pitch of the reporters' voices in Baghdad and Tel Aviv rise with the increasing note of the air raid sirens. It felt real. It was not—not for those of us many thousands of miles away and seeing it only on a television screen. Christopher Coker theorizes that in Desert Storm the grainy images of targeting information, contrasted with the computerized war games used for entertainment, distanced the public emotionally and psychologically. "Here was no community of fate with the enemy," he says.[29] As mentioned above, we saw it without experiencing it. We have assumed the same detachment is true for those actually doing the killing.

Psychologists can claim that killing over distance emotionally and morally disengages the combatant.[30] It is easy to understand this concept if one considers, for instance, a battleship firing its huge guns over the distance of many miles and seeing only the wisps of smoke that cannot translate the destruction and suffering beneath it. When we view weapon impacts on the news, we rarely see casualties. It leads some to question whether we can understand the consequences. One author says, "The question is whether a reliable experience of the consequences of one's actions can be made with regard to these remote-controlled (or even autonomous) weapon systems."[31] The same questions often surface about bomber or fighter crews looking only through cockpit displays.

To take it one step farther, this virtual war (perhaps nowhere as pronounced as it is for unmanned aircraft operators) may remove certain feelings and emotions normally associated with combat. Coker explains that virtual reality cannot simulate the feeling of flying into a barrage of fire: "It cannot reproduce the battlefield experience: the knowledge that in the real world, weapons kill. A number of pilots after a bombing outside Baghdad in 1998 complained of experiencing fear."[32] If simulation could produce surprise in a pilot over experiencing what most would consider a normal response to people trying to kill him, it might be having similar, and probably unknown, effects on remote warriors or those tasked in the future with handling autonomous weapons. Rev. G. Simon Harak, director of the Center for Peacemaking at Marquette University, is concerned about the latter: "Effectively, what these remote control robots are doing is removing people farther and farther from the consequences of their actions."[33] But I suspect there is something missing here.

Grossman details an excellent vignette on the ease of killing large numbers of people from the air during World War II. Tens of thousands of civilians were killed in July 1943 when the Allies firebombed the German city of Hamburg. Grossman says, "If bomber crew members had had to turn a flamethrower on each one of the seventy thousand women and children . . . the awfulness and trauma inherent in the act would have been of such a magnitude that it simply would not have happened."[34] The bomber crews were nowhere close to their victims that night. They could not see individual deaths, but here I think we must return to the words of the Predator wing commander above. Remote warriors, at least in the present form, are not analogous to Allied aircrews. As the colonel said, they sit eighteen inches away from seeing the results of their actions. This could have an effect we have not yet fully characterized. In fact, it is interesting how on the one hand we talk of the sanitized nature of this virtual war, while on the other we sometimes note the severe reaction the public has to seeing the images on television, particularly when things do not go exactly as planned.[35] Perhaps virtual war is not as sanitized as some critics think.

But watching antiaircraft fire on screen or seeing the columns of smoke rising from a recently destroyed target is hardly the same as watching men die. A former deputy commandant of the USAF Weapons School was one of the first close air support fighters to arrive on scene during Operation Anaconda, the botched attempt by U.S. and allied forces to destroy al Qaeda and Taliban forces in Afghanistan in early 2002.[36] Orbiting in an F-16 over what would become known as Robert's Ridge, he saw men shot, one by one, as they attempted to egress a stricken Chinook helicopter. In telling his story to a group of hardened Weapons School instructors and students, he could not keep it together as he explained the events shown on a recording of his sensor display screen. Though he couldn't save all the soldiers, he was instrumental in saving several, at times dropping five-hundred-pound bombs within fifty meters of friendly troops, an act requiring extreme flying skill and courage. He was awarded the Silver Star for his actions. It did not matter to him; he would have traded it to be able to see those guys he saw slain able to walk down the steps of an airplane that brought them home to their families. It is safe to say there are plenty of times, even from the view through a video screen, when the consequences of war are seared into memories forever.

Researcher Maryann Cusimano Love notes military Predator pilots seem to be getting post-traumatic stress disorder (PTSD) at higher rates than soldiers in combat zones.[37] There are anecdotes, like the one above from the wing commander, that tell the story of the long hours of boredom, the exhilaration of doing a job well, and the later effects as the consequences sink in. Experience in performing similar missions from the cockpit of a modern fighter says there must be times when, as the order is given to fire, there is a certainty about the unintended civilian deaths that are sure to follow. We use official, sanitized phrases such as "collateral damage" to ease the conscience, perhaps to allow ourselves to push the button, but those are just words. They fade over a time period much shorter than the images seemingly locked in our minds. There are opportunities to understand the consequences of actions even in virtual war. There is certainly a cost, though we really do not yet know what it is. We cannot now conceive how robotic or autonomous weapons will play into this dimension, though we need smart people to start thinking about it.

Though I have attempted to shed light from a different angle on the psychology of virtual war, a word of caution is in order. Previously I suggested that remote warriors may not be warriors at all. Before that, I brought up the possibility they might not even be combatants. It is important not to misconstrue feelings of regret, possible instances of PTSD, and whatever psychological toll virtual war may take on those who participate in it as evidence of, or even substitutes for, the concepts of combatant and warrior. Emotional connection to the consequences of one's work does not equate to a warrior ethos (though certainly many UAS operators do have one), and it does not solve the deep issue of moral equality that grants the legal status of combatant. Arguing otherwise is to retreat to shallow justifications of status in the frictional world of service culture. This ultimately clouds the fundamental issues at hand. We cannot allow such retreat if we truly hope to solve these issues at any meaningful level below simple pride. We should not ignore the possibility that such feelings or rates of PTSD, if they exist, might be a symptom of a lack of warrior ethos in our military or a subconscious guilt about killing with impunity. But make no mistake: we are asking young men and women to engage in this deadly serious game, one with true if not yet understood consequences, and we owe it to them to ensure that what we are asking them to do is on solid moral footing and in the law of war and that we understand how to care for them when they are no longer involved in this kind of lethal action. This is an area begging for further study, so I leave it to those far better qualified than I to do so.

We move now to another interesting effect of remote warfare, that of living in two worlds. I once talked to a B-2 pilot who had flown combat sorties during Operation Allied Force. During the Kosovo War, every B-2 sortie was flown all the way from Whiteman Air Force Base, Missouri. It made for some long sorties, but it also was a shift in the way Americans went to war. I asked him what it was like to step away from his house knowing in twenty-four to thirty-six hours he would be in combat. He said, "It was surreal."[38] One day he was dropping bombs on Serbia, the next he was cutting his grass. It might be even stranger for those flying UASs on combat sorties from Nevada.

Love quotes a Predator operator saying, "You're going to war for twelve hours, shooting weapons at targets, directing kills on enemy combatants. And then you get in the car and . . . within 20 minutes you're sitting at the dinner table talking to your kids about their homework."[39] I cannot imagine the constant yo-yo of emotion between gearing up for the business of death and winding down so as not to bring it home. Personally, a day or two before I deployed I began a process of withdrawal, a disengagement, from my normal life to avoid the rapid crash that would surely occur when goodbyes are said and the reality of war begins to set in. Is a UAS operator's emotional rollercoaster tempered by the fact he will not, unless catastrophe of another kind strikes, face death while sitting "in combat" in the ground station? Possibly, but his view of death day in and day out juxtaposed against his experience of working on homework with his kids in the periods in between has to take a toll on his psyche.

And what of that issue of the possibility of dying "in combat?" Recall earlier discussions about the legality of killing in warfare, particularly for uniformed military personnel. Whether actively engaged in combat or not, with a few exceptions military personnel are considered combatants and can be killed. This is why I found Walzer's discussions about those soldiers who elected not to kill oblivious enemy soldiers so compelling. They were legal targets even though not engaged at that very moment in hostilities, and the soldiers who passed on slaying them at the time recognized they acted contrary to their duty. O'Connell's concerns about targeted killings taking place outside of her notion of the defined battlefield ought to be our concern as well, though for a slightly different reason.

Others have discussed in great detail the likely possibility that UAS operators, as legal targets regardless of where they are, will be felled in cities right here in the United States in the not so distant future. There is no reason to assume plans are not already under way to target UAS operators, as well as other military personnel, on American soil. There can be no logical counter to the justification of al Qaeda operatives, so long as they follow other laws of war,[40] in doing so. I suspect the average citizen and likely the families of those lost would consider such an act akin to murder or even terrorism, but by the laws of war—in fact specifically because of the laws of war—it would be morally and legally justifiable combat killing. This is not likely to be the biggest concern of those engaged in remote warfare. Many RPA operators, at this still early stage in its evolution, are experienced aviators from other weapon systems. Many have dealt with the abstract knowledge of the possibility of their own death in combat before. More pressing, I suspect, is how their job and its possible consequences affect their families or the other innocent civilians around them.

Retired major general Charles Dunlap, the Air Force's former deputy judge advocate general, writes that as societies become more technologically integrated and dependent on technology, the harder it becomes to separate civilian and military facilities.[41] From the perspective of our own use of force, this means although we have the best of intentions, we will subject more civilians to harm under double effect even with our technologically superior weapons. Electrical grids powering a state's air defense systems may be legal targets, but destroying them would affect thousands of innocent noncombatants. This gets particularly sticky in so-called cyber warfare when attacks may be launched through unknowing third-party states that happen to house a particular computer "server farm." For the case of remote warfare, what Dunlap talks about is precisely what fuels the risk inversion discussed previously. Are not the ground stations housed at bases in the continental United States legitimate targets? Is not the RPA operator driving to or from work a legitimate target? What if that operator is dropping her husband off at work and then running by the day-care center to drop off her children when the rocket-propelled grenade tears through the passenger door and rips her minivan to shreds? Is that a combat death and three deaths chalked up to justifiable collateral damage, or is it the bloodthirsty act of a crazed ideologue?[42] Well, it depends a lot on perspective, I suppose. That is, unless you understand the laws of war.

It is here on the Las Vegas strip, as tourists gawk in awe at the fountains of Bellagio and sway to the piercing melody of "Con Te Partirò," where the desire for warfare with impunity collides head on with its ironic result. In attempting to "fight them over there,"[43] farther and farther from the home shores, we have turned our own cities into legitimate battlefields. As Reverend Dr. Harak says, we have brought the war to our shores as enemies will surely adapt and begin to attack our "command centers [notwithstanding the attack on the Pentagon on September 11, 2001]. . . . The whole notion that we can be invulnerable is just a delusion."[44]

There is more than meets the eye with regard to the CNN effect. For the case of remote warfare such as is conducted by RPA operators, there is no doubt the killing can feel very real, perhaps more so than for fighter or bomber crews who never see the aftermath of their actions with such clarity. The fact they kill with impunity and have yet to be confronted with the reality of their own death in combat has to have some psychological effect, but we just do not know yet what it is or even how to characterize it. It is possible RPA operators will face greater instances of psychological trauma due to their actions, but it is as likely much of this postulated trauma occurs because of the constant swinging back and forth from "combat" to home life. This is a new way to fight a war, and there is simply no way to predict how this combination of killing with impunity mixed with the emotional pendulum of normal residential life will turn out over the long term. We need a concerted effort to discover the intricacies and inform future decisions on unmanned and robotic warfare and the best methods to employ it.

Finally, there are severe societal consequences to fighting a "global war" from the American heartland. As others have warned, we have opened ourselves up to a cold irony and provided justification for those we would rather be fighting somewhere else to bring the fight right to Main Street. How does this play on the mind of our remote warriors, and what effect will it have on the culture at large? We simply do not know. Unfortunately, I suspect we are all going to find out eventually.

Wanted: Renaissance Men and Women, Understanding of War Required

In early January 2007, I was directed to attend a meeting at the Rand Corporation's offices in Arlington, Virginia. Apparently as part of my duties at Headquarters United States Air Forces Europe, I was to join a group tasked with solving the aircrew management problem. It was a daunting task; we had never had it exactly right for as long as there has been a personnel system attempting to manage aircrews. The group was called the Transformational

Aircrew Management Initiative for the 21st Century and was going by the unfortunate acronym of TAMI 21. It was chartered in October 2006, and its first plan did not survive first contact with the chief of staff of the Air Force. Little did I know at the time, the proceedings of the working group would foreshadow the organizational structure and personnel issues inherent in any service's move toward unmanned systems and autonomy.

One of TAMI 21's proposals was to reduce pilot production, particularly that of fighter, attack, and reconnaissance pilots. That did not sit well with the chief of staff, and in December 2006, following the initial TAMI 21 brief, he nonconcurred on most of the proposals and mandated a ramp-up of pilot production over the coming years.Unfortunately the pilot training pipeline was incapable of producing such numbers, and the squadrons were incapable of absorbing them.[45]

Manning was a complex problem due to closure of several F-15C units in preparation for the F-22, an aircraft not yet open to pilots fresh from pilot training. Therefore it required already experienced fighter pilots to fill its cockpits. The closure of five F-16 squadrons along with the reallocation of their aircraft to other units by the Base Realignment and Closure program for 2005 added to fighter manning problems.[46] Additionally, in the haste to draw the entire Air Force down to congressionally mandated end-strength numbers, Program Budget Decision 720 cut already undermanned maintenance units even further,[47] which reduced their ability to generate the numbers of sorties required to keep the growing overage of pilots trained. It was a vicious cycle of overmanning in squadrons, a fixed number of flying hours allocated per month regardless of the number of pilots, and a maintenance organization unable to produce more hours even if more could be allocated. These issues, while complex, paled in comparison to the personnel problem that parallels our future move toward robotics.

At the same time these drawdowns took place, there was an explosion in "requirements" for pilots with fighter experience on staffs, in war-fighting headquarters, and in air operations centers (AOCs) all over the world. These billets were for mid-level officers—majors and lieutenant colonels—with years of experience in fighter cockpits. All in all, there was a stated need for approximately four times the number of experienced fighter aviators in these kinds of billets than the Air Force was capable of producing in new pilots every year.[48] This was an unsustainable inverted pyramid of personnel requirements. This issue mirrors a looming problem in the acquisition of robotic systems. It is a good thing the Rand reports on the aircrew management problem—and there are several—are still available. With the advent of autonomous warfare, we are going to need to dig them out to aid in solving the issues of military deskilling and growing the right kind of generals.

In science fiction, there seem to be two well-established camps on how to view robots. One is the apocalyptic version of *Terminator* and *I, Robot*. The other is the utopian view whereby they are our everlasting helpers, such as the Data character of *Star Trek: The Next Generation*. Similar views hold for thoughts of future military robots, with various factions inside and outside the military coming down on both sides. The DOD's *Unmanned Systems Roadmap* is on the utopian side, as are the major manufacturers and trade representative organizations. How could they be otherwise?

As I researched this work, the overwhelming response to my question of whether robotics manufacturers, industry organizations, governmental program offices, or research agencies are concerned with the ethics of robotics was a simple "No."[49] When I asked a DARPA science, engineering, and technology assistant (SETA) if in his time at DARPA he had ever heard of a proposal for an ethics review board—one made up of ethicists, philosophers, engineers, and others with no decision authority but who could advise on possible ethical issues and consequences of emerging technologies—he said, "That doesn't seem like something DARPA would be concerned with."[50] This is completely understandable and not at all sinister. DARPA exists to push the technological envelope. Its mission is "to maintain the technological superiority of the U.S. military and prevent technological surprise from harming our national security by sponsoring revolutionary, high-payoff research bridging the gap between fundamental discoveries and their military use."[51] It does not really ask the "should" questions. Those things are for others to work out. At DARPA, other research agencies, and in the minds of most inventors and engineers, these things are just wicked problems to be solved. The question is simply "Can we make this work?" It is often only after the fact, as with some of the Manhattan Project scientists, that conscience tends to set in.[52] In fact, in my experience among trade representatives, the general response to dissent or simply questioning whether some things ought to be pursued is along the lines of "it's going to happen; it's just a matter of time." The implication is "you better get on board." It turns out that stifling debate, or just not pursuing it rigorously enough, often leads to disaster. Examples abound. So, if we cannot talk ethics in these fora, maybe we can talk practicality.

In another utopian view of robotics, one officer implies in an Army War College thesis that he sees the coming robot army as a chance for mediocre officers to have their day in the sun. He does not specifically say so, but he does say "commanders will not have to be as good at motivating their soldiers. They will only have to focus on operations, not on the human element of conflict." And since there will be less friction, the "commanders do not need to be as good."[53] Ignore for a moment the fact that technological revolutions actually create more friction and disruption in everything from organization to culture to concepts of operations, and that a reduction in Clausewitz's friction has been an oft-dreamed, never-realized idea throughout history.[54] As I allow time to think about the implications of the officer's statement, I'll pose a very basic question: how will that commander learn about the operations he is to focus on?

If the control systems that people such as Missy Cummings are working on come to pass and we have swarms of unmanned vehicles in all domains operating at some level of supervised autonomy, the man on the loop will need a vastly different set of skills than someone who has done nothing more than drive a PakBot around, send out a USV to check out a suspected drug runner, or fly a Grey Eagle (the Army's Predator-like drone) for hours on end while staring at the same piece of dirt or the same suspected insurgent. In the not-too-distant future, how will a land commander direct autonomous vehicles from a command center and understand what it is they ought to be doing if he has never been out there riding on similar ground and testing his skills, actually out in the elements and over the terrain, in places such as the National Training Center in California? How will a joint or coalition air component commander understand the intricacies of the air operations plan he is responsible for executing if he has never led or at least been a part of large-force missions in training exercises such as Red Flag in Nevada or Alaska? How do we expect to season the future leaders of robotic warfare?

Lots of plans look fine on paper or in the computer code of advanced simulations, but many fall flat when exposed to the multivariable nature of the real world. Out there is where the ultraviolet rays of the sun beat down, rainwater runs into everything, hydraulics freeze, nothing escapes the ages-old process of oxidation, sand stops moving parts, and the ever-fertile furrows of the human mind make adversaries do things no one ever expects. There is a complex problem ahead involving the deskilling of our force, the problem of growing generals without the benefit of experiences of the past, and the unsettling possibility that military control over lethal force might someday fall to "lab-coat warriors" with extensive knowledge of robotic engineering but scant knowledge of the operational art of war.

During Operation Iraqi Freedom one of my former wing commanders was in charge on the Combined AOC (CAOC) floor when a key communications antenna in downtown Baghdad was on the strike list for the day. It was an important target because it allowed the former regime to continue directing certain loyal elements in action against coalition forces. The general described being in direct communication with the Predator crew as he watched the feed on one of the many CAOC video screens. He directed the placement of the weapon aim-point cursors on the antenna. He was not actually controlling the sensor, but he told the sensor operator where to move the cursors to ensure destruction—a little left, up a bit—then he gave the authorization to fire.[55]

For those not versed in Air Force doctrine or the history of command and control of air forces, this was an unprecedented departure from both. The Air Force still claims the concept of "centralized control, decentralized execution" in doctrine.[56] This means a central air operations plan sets out the operational goals and objectives, but tactical execution is left to lower-echelon commanders and, ultimately, to the crews responsible for employment. A general, responsible for the overall operational plan, directing individual weapon placement and giving clearance to fire is well outside the bounds of the "centralized control, decentralized execution" mantra. This trend toward *centralized* execution is what Peter Singer describes as creating "tactical generals."[57] Unfortunately for those who believe general officers should concern themselves with operational and strategic-level ruminations, it has been going on for a long time.[58]

The no-fly-zone operations of the eleven years leading up to Operation Iraqi Freedom and the growth of the concept of the air operations center as an actual war-fighting element—one Air Force chief of staff even called it a weapon system in and of itself—gave plenty of opportunity for leaders to remain in the tactical weeds long after they should have left it to the captains. During Operation Southern Watch, it was not uncommon to hear the AOC director order individual aircraft or flights around in Iraqi airspace. The problem with centralized execution is that it takes a level of understanding that will be increasingly difficult to come by the longer it persists due to atrophy of skills at lower echelons and a perfection of available data that does not exist and perhaps never will. As Singer points out, there is latency in the information presented to the generals in command centers, and often not all of the information is displayed—for instance, the exact numbers and locations of subordinate units.[59] Such "reach through" by generals into the tactical realm can produce errors, but their impact on the organization is of far more consequence.

Two questions come to mind. If the generals are doing the captains' work, who is doing the generals' work?[60] More fundamental to the future of military organizations is, who is teaching the captains to be generals? As the rise of information technology has "flattened" the world, to borrow the term Thomas Friedman made famous, the military trend has been toward steeper vertical lines. The temptation to dive into the tactical level has proven too hard to resist given the capability to do so. The danger is as more and more decisions are made at higher and higher levels, we never allow our young officers the value of the school of hard knocks. Over the course of the last twenty years, there has been a slow migration of the ability—authority—to make real-time tactical decisions in fighter and bomber cockpits. These decisions are often directed from above. It brings to mind a thought of George Orwell: "The greatest mistake is to imagine that the human being is an autonomous individual. . . . Your thoughts are never entirely your own."[61] With a loss of a sense of autonomy there is a natural tendency to absolve oneself of some of the responsibility as well. Why take responsibility when you are robbed of authority? These are not good trends for men and women who still have their fingers on the triggers. Eventually, if taken to the extreme, we will have to promote those who have mostly been told what to do and how to do it into positions where it is now their job to do the telling. How will they respond? No one can possibly know.

I must highlight there are certainly exceptions to this. Driven mostly by our focus on counterinsurgency operations of late, junior officers may be in charge of small towns or large areas and may be responsible for everything from security to electricity and water. They are essentially the department of public works, and there is no one there to tell them how to do it. I suspect this is not a by-choice decision of higher-ups in the Army and Marine units primarily executing these operations; the quantity of information is simply not available at higher levels. One has to ask what would happen if it were, but clearly these young officers are gaining valuable experience in decision making that will serve them and the nation well in the coming years.

As they relate to unmanned or autonomous warfare, the issues surrounding the ability to grow leaders are really twofold. First, there is the problem of raw numbers. In a world of supervised autonomy or full autonomy, there is not a need for the number of lower-level operators, and this shrinks the available pool for promotion and greater responsibility. When you are in the market for "renaissance men"[62] and women with a broad knowledge of operations, you would like to have more to choose from. The other issue is being able to catch the trend early enough in order to address it.

The Air Force's *Unmanned Aircraft Systems Flight Plan* addresses the numbers and skills issues. This is encouraging, but it seems more concerned with career advancement opportunity than the issue at the heart of the matter—having the right people in advanced ranks who have a broad understanding of operational art and the systems they will employ. Still, a recognition of the challenges for future leaders already spelled out in a vision document means we are on the right track to addressing the problems. The *Flight Plan* says:

> *A key challenge to realizing the vision will be to develop and maintain the right skill sets of systems and operational software developers, mission directors, and future USAF leaders.* . . . The team members need to be selected for basic skills and then further trained to build systems that can *fight the battles at all levels of conflict* in all environments. *Relatively few mission directors will be needed so issues of career advancement and selection criteria will be challenges for future leaders.* These leaders will also require different skills to employ air power that is largely non-human. In the future, the warrior will have incredible combat power and responsibility. [emphasis added][63]

The DOD's *Unmanned Systems Roadmap* is more explicit on the issue of deskilling and the challenge posed by the removal of a level of operators whose job it is to live in the tactics and gain experience for their later jobs as operational, possibly strategic leaders:

> The automating of the actual operation/fighting of platforms will decrease the need for people to crew them, while the personnel needed to simply maintain the vehicles is likely to increase. This has the potential to radically change the "tooth to tail" personnel ratio in combat forces heavily in favor of the support personnel vice combatants, increasing the need for resources (people, equipment, money) for the support chain. At the same time, the need for experienced middle to senior combatant leaders and decision makers will not change, since they will know the tactics and strategies necessary to operate and direct the autonomous systems. *The challenge will be developing the middle to senior combatant leaders needed in an environment allowing fewer junior combatant leaders.* [emphasis added][64]

By recognizing these issues now, the DOD and the Air Force are also addressing the second problem, that of catching the trend early enough.

The man-on-the-loop concept and the idea that UASs are being developed to support other manned aircraft operations is probably not publicized enough. Other services do a better job of discussing their vision of robotics as enablers of, and adjuncts to, manned systems and the soldiers and sailors they will support.[65] If Air Force recruiting commercials are the only thing used to judge the future of the flying force, it appears we will be flying nothing but UASs. This is clearly not the case, but for some reason the Air Force seems reluctant to discuss this vision and ideas on the levels of control. The man-on-the-loop concept specifically addresses this and is a hedge against warnings that we will be caught unaware with respect to abdicating human decision making. Consider what political-military strategist Thomas Adams says in *Parameters*. He warns it will not be a conscious decision to remove ourselves from the loop, but we will become more and more irrelevant by moving "incrementally toward systems whose logic demands that human control become more abstract with less direct participation."[66] There are ways to mitigate it—the *Flight Plan* at least recognizes the issue—but it is a real danger. This technology's insidious nature is a running theme in this research.[67] As there is less human participation in decision making, there is also the danger of losing certain skills altogether.

Krishnan says, "Automating warfare and using military robots for a great variety of functions effectively devalues traditional military skills and could lead in the long term to the complete extinction of the military as a unique or at least distinct profession."[68] That is a particularly bleak view, but it ought to be of concern to both military professionals and the public they defend. There is a concern that once autonomy is advanced enough, teams of robots, particularly those programmed with learning autonomy, could go forth and complete missions without human intervention. As this capability propagates, there will be a skills transference from the traditional military member, versed in tactics in an early career and grown into an understanding of the operational art later in life, to those software developers mentioned in the *Unmanned Aircraft Systems Flight Plan* who need not understand "any traditional military skills or virtues."[69] Because automation and autonomy will likely happen first at lower levels then move upward, the skills and knowledge required of those at the top will grow while those at the lower levels eventually become redundant.[70] This creates the same kind of inverted pyramid the TAMI 21 group faced with aviators in the late 2000s and the *Flight Plan* addresses above. TAMI 21 attempted to fix this issue by addressing bogus requirements.[71] No such solution exists for this issue. The *Flight Plan* is right: one of the major issues facing military leaders will be how to train those who will replace them in a world of autonomous warfare.

But there is another consequence of deskilling that should be of greater concern to the population at large. The *Flight Plan* mentions a "team" of software developers and mission directors joining to create systems that can fight on many levels of conflict. It may be a new paradigm for combatants in an age of autonomous warfare, but it could fundamentally alter our understanding and requirement for military control over lethal force in war. In discussing the issue of responsibility and accountability for lethal robotic systems, Ronald Arkin argues that the decision to use lethal force will always remain with a military commander. In his view, lethal robots will be programmed for a task, the military personnel responsible for the robots will "sign off" on constraints in place to conform to current ROE (thereby completing the accountability chain), and the robots will be sent off to do the dirty work.[72] It sounds good in theory; however, there is a possible regression whereby military control becomes a mere façade.

As this proposed team of engineers and "mission directors" morph with greater advances in learning autonomy and the only real skills required become technical in nature—maintaining and programming robots—it is possible for "military control" to be exercised by individuals with virtually no military training whatsoever, as Krishnan warns. There is a possibility that an engineer, no different than a civilian engineer employed by the company contracted to provide the robot, will authorize lethal force under the simple idea that he is wearing a uniform.

The services have employed engineers for a long time, and they do great things for the nation, but they have never been the ones authorizing or employing the use of force. Those design teams that worked on Tomahawk and Maverick missiles, systems portending the age of robot warfare, were not the people who fired them off of ships or aircraft missile rails. We have always employed a different group—dare I say, a warrior class—to take the weapon the last "tactical mile" to the target. If we are serious about maintaining the true nature of military control over lethal force in war, we will consider the ramifications of our desires for increased use of lethal unmanned and robotic systems very carefully.

Ethos, Lethal Force, and the Public Trust

There are tangible practical concerns about proceeding farther down the path to robotic warfare. Senior military leaders will have to confront the training and education issues involved in the possibility of military deskilling and how it may affect future organizational structure. We will need to ask ourselves how we intend to grow future senior leaders with the broad skills to control the madness of warfare when the training ground of direct tactical experience shrinks away. We should be concerned with the trend toward more vertical decision-making structures in the military, driven solely by the ability to do so, while we see the rest of the world dealing with the information revolution in the opposite manner.

Finally, if the military is to be a good steward of its capability to use lethal force and maintain the public trust to do so, it must confront head-on the possibility of an insidious slide away from the true intent of military control over such force. It cannot allow itself to be seduced by the technology to the point it lets go of its objectivity. The military, and the public it defends, ought to be wary of the hand-waving and jump-on-board message of industry and those organizations that represent them. Crushing dissent inside the armed services "for the overall good of the Joint Force," as the *Unmanned Systems Roadmap* states, is one thing.[73] That is just bad policy. Should we end up shielding the public from such debate due to a desire to procure these capabilities, it would be, quite frankly, a dereliction of duty.

Such a debate on exactly what military control over lethal force *is* ultimately boils down to what form the public thinks its combatants ought to take. We currently have a warrior class, those at the tip of the spear and able to employ force in combat. Whether we have, or can maintain, a warrior ethos is debatable. I submit that it is desirable in that it sequesters, in a small number of highly trained and educated individuals, the necessary mental and technical skills required to properly—justly—employ force with the intent and consequence of destroying human life. These are concepts that do not come down to simple ones and zeroes. They do not come down to donning a particular uniform or reciting some creed. A concerted effort to maintain such an ethos does not necessarily preclude advancements in unmanned and robotic warfare, though there are significant hurdles to overcome. It is conceivable such an ethos may even enhance our resilience and ability to deal with the warrior dimension of impunity in warfare. Regardless, we cannot allow it to atrophy without due consideration and without something to replace it. As Charles Dunlap warns, "although the extent to which the proliferation of long-distance, push-button war serves to replace that ethos with a new ethic is as yet uncertain, it is imperative that whatever emerges instills in tomorrow's soldiers those moral underpinnings which will further develop the application of ethical and legal norms in future conflicts."[74] I could not agree more.

7

Impunity and the Future of War

We're talking to them in a way they can understand.

—President George W. Bush, on the efficacy of targeted killing[1]

A few years ago I was sitting around a table with my fellow squadron commanders discussing the futures of many of our charges. It was assignment season, and the difficult questions of who should go where doing what were hanging over the meeting. In these sessions, dreams were helped along for some younger pilots and crushed for others. Most captains flying fighter jets saw themselves doing nothing else, and they were uninterested in the myriad other "opportunities" the Air Force was about to offer the majority of them. Only 20 percent or so would be seeing another tour in a fighter after their current assignment.

We were discussing new platforms such as the RC-12, a modified Beechcraft Super King Air, being used for surveillance and reconnaissance in Central Command's area of responsibility, and the Air Force's proposed light attack aircraft, to be used for training other nations' nascent air forces in platforms less complex than fourth-generation fighters. Our commander threw an article on the table with a picture of a Cessna Caravan—a single-engine, turboprop aircraft often used for light cargo or passenger movement by small regional airlines—being tested at the Navy's China Lake facility in California. It was firing a Hellfire missile and was slated to be delivered to the Iraqi Air Force when testing was complete. There were a few advisory jobs for young aviators—really pilot training positions—in such places like Kirkuk, where the new Iraqi Air Force was training.

In seeing the photo of a distinctly civilian-looking aircraft firing off a Hellfire, I was reminded of the variety of aircraft the Air Force had operated during and immediately after the Vietnam War. Twin-engine, propeller-driven aircraft such as the OV-10 Bronco and O-2 Skymaster were "low-tech" planes suited for the kind of war we found ourselves in back then. They truly were light attack aircraft capable of firing rockets and, in the OV-10's case, small-caliber guns and even air-to-air missiles.[2] I jokingly remarked, "Welcome to the Air Force of the past—I mean future." It was met with chuckles from my peers and a grimace from my boss. Despite the lighthearted editorializing, there is no denying the underlying truth. Boeing even considered bringing the OV-10 back to vie for the light attack aircraft concept contract.[3] It is very possible the U.S. military of the future is indeed one of the past.

War of the Future—and Past

Here, more than twenty years since the fall of the Berlin Wall and the implosion of the Soviet bloc, we might just be getting a peek at what the future of warfare will bring. For at least the next several years, possibly a decade or two, the U.S. military is unlikely to face a conventional force like the one it faced during the Cold War. Former secretary of defense Robert Gates stated during a speech at West Point in 2011, "The prospects for another head-on clash of large mechanized land armies seem less likely." He followed by saying, "The odds of repeating another Afghanistan or Iraq—invading, pacifying, and administering a large third-world country—may be low," but the Army and the rest of the government must focus on capabilities that can "prevent festering problems from growing into full-blown crises which require costly—and controversial—large-scale American military intervention."[4]

This concept of preventing "festering problems" likely means the U.S. military will find itself in locations all over the world, deployed there in much smaller footprints, working to prevent large wars from occurring but facing multiple small-scale threats along the way. This, as I alluded to before, is what the former chief of staff of the Army, Gen. George W. Casey Jr., called "persistent conflict":

> We believe the global trends portend several decades of persistent conflict in which local and regional frictions, fueled by globalization and other emerging trends, are exploited by extremists to support their efforts to destroy our way of life. . . . Future conflicts will occur in many forms as violence ebbs and flows across the spectrum of conflict—ranging from

> stable peace to general war and all points in between—and in each of the physical domains—in the air, at sea and on land.[5]

This view is not under the sole proprietorship of the U.S. defense establishment. British Army general Rupert Smith, former commander of UN forces in Kosovo and former deputy supreme commander of NATO forces in Europe, writes in his book *The Utility of Force: The Art of War in the Modern World* about the trend toward "war amongst the people."[6] He defines this kind of war "as a world of confrontations and conflicts" that has been around for centuries but "became the dominant form of war at the end of the Cold War."[7] This kind of war is manifest in places all over the world, most notably in Iraq, Afghanistan, the Democratic Republic of the Congo, and in the Palestinian territories, but it is not limited to them.[8] Almost anywhere you find a failed or failing state, you will find conflict of the kind both Casey and Smith describe.

Smith distills six trends for war among the people. All are thought-provoking, but three have direct bearing on the future of war and unmanned and robotic systems' place in it. The first of these trends reinforces General Casey's contention: Smith says these "conflicts tend to be timeless, even unending."[9] This is bad news for a military stretched thin and tired from ten to twenty years of combat operations tempo. Second, the fight is among "the people," not on the battlefield. The concept of an actual battlefield may be fading into history, though one could argue that the nearly total war of World War II made it a vanquished historical concept many decades ago. Third, "we fight so as to preserve the force rather than risking all to gain the objective."[10] In fact, we may not even know the objective and instead often fight to establish "conditions in which the outcome may be decided."[11] These three trends are sure to exert significant pressures on the continued vector toward unmanned and robotic systems, but Western ideas about technology in war will likely clash with the implications these trends may herald.

I have examined at length the desire to fight so as to preserve the force. This is the rationale for our low-risk form of war, now approaching impunity. Clearly this drives the desire for automation in warfare from a Western perspective, but as I have discussed, it has profound effects on what it means to be a warrior or even a combatant. I will leave the subject for the reader to ponder those issues already raised as they relate to considering this world of persistent conflict. The war-among-the-people trend is equally troubling for the purveyors of robotic technology and those on the demand side so eager to acquire it. These are wars where individual relationships on the ground tend to matter more than firepower. If Army Field Manual (FM) 3-24, *Counterinsurgency*, is correct, popular support for or tolerance of an insurgency is its center of gravity.[12] That means it must be the primary focus of any counterinsurgency to replace that support or tolerance, to win the "hearts and minds."[13] How does a robot have tea with a village elder? I will leave this to the later section of this chapter on the "conversation" that must take place between combatants in all war, but it is particularly important under this new paradigm. For now, allow me to address the trend of persistence and highlight some issues as they relate to robotics.

For air forces involved in armed conflict for over twenty years and ground forces now surpassing ten years in combat, the grinding pace of deployment and overuse of equipment are strategic concerns. The Air Force is now focused on recapitalizing its fleet of many different platforms. General Casey, in his time as chief of staff of the Army, worked on resetting deployment schedules in order to allow for needed time to train and recuperate while at home.[14] Timeless and unending conflicts scream for solutions that differ from what we are going through today. Granted, much of the operations tempo is based on fighting two major land wars in the Middle East and Asia, and Secretary Gates's concerns about doing so again are well founded. Automation serves to reduce the stresses on human capital, and though we most often consider this effect in industry, earlier discussions about unmanned vehicles partially centered on the manning requirement. There was a sense this could be a way to solve some of the manning shortfalls or increase capabilities even while drawing down the force due to mandated end-strength policy. Such were the issues that drove such decisions as Presidential Budget Decision 720, discussed previously. These euphoric, perhaps utopian thoughts have now encountered reality.

Manpower requirements for operating unmanned systems sometimes exceed those of manned systems.[15] The *Unmanned Systems Roadmap* recognizes this issue and anticipated the questions that are sure to come from members of Congress, the holders of the purse strings, who do not understand the intricacies of fielding unmanned and robotic systems. They are likely to ask the tough fiscal questions about exactly what kind of manpower it takes to deploy unmanned systems. Eventually the hope will be for single operators to supervise multiple unmanned systems. This is not possible today, though it might be soon. Such a capability may again alter manning requirements and create more turmoil. For now it seems clear there is an increased need for the logistical support. The *Roadmap* says the need for personnel to maintain and service these systems will likely increase, and this will drive a greater "need for resources (people, equipment, money) for the support chain," as I noted previously.[16]

Here is one current, real-world example. When iRobot successfully got its PackBot into the combat theater—recall it was a DARPA technology demonstrator whose problem had finally arrived—and into the hands of twentysomething soldiers (particularly those involved in disposing of explosive ordnance and defusing IEDs), demand grew for the system. Therefore, demand grew for being able to maintain or replace the system. That job fell to the Robotic Systems Joint Program Office (RS JPO), part of the OSD umbrella of the joint robotics program, based in Detroit where it set up a "depot maintenance" facility at Selfridge Air National Guard Base.[17] Though they are able to receive a damaged robot from the theater in four days, fix it within four to eight hours (or write it off and replace it), and get it back into theater within the next four to six days, there was a need for a faster turnaround in theater on more minor repair issues. The RS JPO set up six locations in CENTCOM's area of responsibility for this task, with two more on the books, and manned it with soldiers who volunteered for additional training in robot repair.

It sounds like a perfect system, but for personnel management professionals there is one glaring problem. At the time, the RS JPO managed five systems; only two were programs of record. That means only two came with sustainment funding and life-cycle costs. The others, and the majority of the work done by the Selfridge repair facility, was funded by Overseas Contingency Operations funding, a special category of money used primarily to fund operations in Iraq and Afghanistan. It means there was no impetus in the Army to carve out dedicated personnel for this kind of repair work. The volunteers were taken "out of hide," meaning there was no backfill for a unit that gave up a soldier or Marine to do this work. There was also no separate career field for robot mechanics; there was not even what the Army calls an "additional skills indicator" for those who have gone through the training for repairing robots. This means the Army had no institutional means of tracking those soldiers with this skill and therefore could not effectively manage training needs across the force or personnel placement for units requiring this capability. This example shows we are already seeing the bulge of the tail in the "tooth-to-tail ratio," and we are dealing with it the way we always have since someone first coined the term "more with less," or at least more with the same. We are doing it with a can-do attitude, abundant but temporary funding, and a consequences-be-damned mentality that we can just figure it all out later.

There are no simple solutions to these issues. In a time of persistent conflict, hard choices will have to be made about the makeup of the force. In the current fiscal environment, increasing manpower is a nonstarter. This problem of finding the right mix in the force is what former secretary Gates was alluding to in a speech at the Air Force Academy when he said the services "must think harder about the entire range of these missions and how to achieve the right balance of capabilities in an era of tight budgets."[18] As the tooth-to-tail ratio bulges toward the tail, as the *Roadmap* expects and the example above proves, which functions will be traded away in order to build a robot-maintenance function and others that will be needed? Is it even the right thing to do based on desired outcomes in wars among the people?

We may desire to automate warfare, but it may not be the right thing to do if we are in fact in a new war paradigm. We will need to make informed choices on the right mix as we proceed into the decades of this new way of war. As Christopher Coker has said, and Secretary Gates seemingly understands, the enemies of the future will be those of the past.[19] What he means is that we will face enemies who do not—who cannot—meet us on the enormous field of battle in tanks and with disciplined infantry forces, with support from large air forces and air defense systems, or in epic naval battles between massive fleets of ships. Rupert Smith provocatively puts it this way: "War no longer exists. Confrontation, conflict and combat undoubtedly exist all around the world . . . and states still have armed forces which they use as a symbol of power. Nonetheless, war as cognitively known to most non-combatants, war as battle in a field between men and machinery, war as a massive deciding event in a dispute in international affairs: such war no longer exists."[20]

We will face small bands of armed men (they will be overwhelmingly men) who, though we will not like to admit it, have the ethos of warriors and consider themselves as such. We will face cultures that understand the existential nature of war in places such as the tribal regions of Pakistan, in Yemen, Rwanda, Somalia, and places all over the globe. For twenty years we, the U.S. military, have been chastised for holding on to the war of the past. The criticism is focused on our large conventional forces built for the possibility of the Cold War turning hot. The irony is there is a possibility we ought to be focused on a past even deeper still. The double irony is that even as we slowly bend toward persistent conflict, we are doing it with our traditional technological leanings in search of asymmetry and impunity, the combination of which might just create a force even more divergent from the war we face in the future than the conventional force we have today is divergent from the war we have in the present.

The Double-Edged Sword of Asymmetry

Martin Cook tells a story about a Naval War College student, an Army officer and former tank commander on the cusp of becoming a senior military leader, discussing the proper response to incoming fire from a hypothetical Iraqi town. The officer's response, in this academic situation, was to return fire with his tanks and level the town. Though I was dubious of the claim and still remain so, thinking we are being kept out of some defining context, Cook claims this was part of a continuing conversation with this individual, whose opinion did not change over time.[21]

The "doctrine" described by the officer is that of massive and overwhelming firepower. Though there are significant concerns about proportionality in this particular example, the concept of overwhelming, even superior firepower is well founded in all military doctrine, but it is particularly prevalent in the U.S. military. We do not take knives to gunfights. The U.S. military and most Western militaries attempting to emulate it have long lived on this concept. It has created a vast asymmetry in military capability and laid the foundation for sweeping victories against other conventional forces. During Operation Iraqi Freedom, coalition air forces faced almost no opposition from the Iraqi Air Force, and in the aftermath of major combat operations teams actually found Iraqi fighters buried in the desert.[22] It was a poetic end to the service whose actions began coalition air forces' efforts to protect the citizenry it had turned on in the tenuous months following Operation Desert Storm. Such superiority in conventional military means drives adversaries to seek asymmetrical advantages. They cannot compete on the basis of pure military power, but they can cause real damage and incite a fracturing of willpower in the public supporting military operations.

The growth of IEDs in Iraq and Afghanistan are tangible signs of the attempt at asymmetrical warfare. The United States has now spent millions on systems to limit their effectiveness. One such system is the M-160 robotic vehicle, one of the two programs of record that the Robotic Systems Joint

Program Office manages, designed to detonate or disable explosives during route clearance operations.[23] Others include electronic jamming devices and the mine-resistant, antipersonnel protective vehicles bought en masse for operations in Iraq and Afghanistan. The IED problem and these programs' attempts to combat it might be indicators of what could occur with the further proliferation of robotic technology.

There is a concern that using robotics could create a global arms race in these technologies or cause our adversaries to react in other asymmetrical ways.[24] We are in a low-grade arms race right now against insurgents, each side trying to outperform the other's technological answers to the tactical problem at hand. Of course, as is our tradition, our solutions are far more complex and costly. Deploying these technologies might also cause others to respond with unpredictable, asymmetrical, and disproportionate force.[25] Coker claims the nation's "attempt to make war more humane for its own soldiers and the viewers back home is making it increasingly vulnerable to the kind of asymmetric strategies we saw . . . on September 11, 2001."[26] It is a lack of capability to respond in kind that may drive such actions by those we are fighting now or will fight in the future.

If a perceived inability to counter a superior military force lies at the heart of the turn toward international terrorism, what will be the effects of deploying combat "droids" to do our dirty work? An enemy, faced with the choice of fighting a force upon which it is incapable of inflicting harm or attempting something else altogether, is likely to deny combat and branch out into other venues. This is the concept feeding the problem of the risk inversion. I believe it is our great fear about terrorist organizations gaining access to weapons of mass destruction. If the virtue of the weapon lies primarily in the character of its user,[27] and we face an adversary whose twisted sense of its ideology sees civilians as legitimate targets, we believe there is no form of restraint to keep them from using it.

But even short of this, there are other concerns. As discussed in the previous chapter, the impunity with which we currently employ unmanned aircraft is not likely to last. Such impunity and the "lack of capacity to respond in kind" by the weaker side currently gives those with the technological advantage a monopoly on targeted killings.[28] This is as you would expect given an understanding of the spectra of impunity in warfare, but what if it were not so? If the other side could begin to target drone pilots wherever they are and no matter what they are doing, Michael Gross claims there might be an impetus to return to a convention prohibiting targeted killings. He then admits that once out of the bottle, "it might not be that easy to get it back inside again."[29] This is only theoretical, but it ought to serve as a warning. The monetary cost of this technology will come down as has been the case for all technologies ever presented to market forces. David Koplow warns, "We must ponder not only how these dynamic new implements might be advantageously wielded by our own troops, but how, over time, they might be turned against us by malefactors including enemy states, terrorists, and street criminals."[30] Even if we can program our autonomous lethal systems to follow the laws of war (discussed in following chapters) or can employ them only in sanitized situations where collateral damage and the concept of double effect are not relevant, others can simply forgo those constraints in uses against us.

What we must understand even if we are moving into a new paradigm of war, one conducted persistently and among the people, is that it does not change the fundamental concerns about weapons or tactics, which remain "anchored in humanitarianism."[31] Gross explains the "crux of every dilemma of asymmetric warfare" is balancing military necessity and humanitarianism.[32] These may actually counter each other in wars among the people. What passes as our military necessity may be seen as wholly disproportionate, as is the case now in the Pakistani tribal regions and in areas in Afghanistan where one civilian death in a drone strike becomes an international incident straining already taut ties between the United States and those two countries' governments. In our quest for the impunity and the military necessity of targeting those enemies we might not otherwise be able to engage were it not for our unmanned aircraft, we may be removing any possibility for our adversaries to respond in ways even partially compliant with the laws of war. The argument that we might be producing more adversaries than we are killing undermines justification of such tactics by the reasonable-chance-of-success criterion of just war. The possibility that we may be legitimizing a response wholly outside the bounds of justification or legality in warfare is far more ominous. It could elicit an equally disproportionate use of force from our adversaries. We may move beyond the current single-shooter tactic we have now witnessed most recently with the attack on USAF Airmen in Frankfurt, on air advisors in Kabul, and in other locations all across Afghanistan. We may unleash a far greater response, particularly if terrorists one day acquire a weapon of mass destruction. As Gross warns, "The enduring dilemma, for both sides, is to wage asymmetric war without undue harm to noncombatants. The doctrine of disproportionate force is not the answer. Disproportionality, by its very nature, knows no restraint. It has no built-in limits that invoke the imperatives of humanitarianism."[33]

Let us hope our tank commander above has reconsidered his position on responding to the report of a single rifle. Let us think hard about the effect of asymmetrical impunity on the enemy in our age of persistent conflict among the people before electing to further deploy lethal unmanned systems. If by our action we leave our adversaries no opportunity to respond in kind, human nature says they will find ways to respond "out of kind." To the extent our way of war gives them tacit legitimacy in doing so, we should prepare for disproportionate force to be turned against us. One good model would be to consider the lot of an innocent noncombatant on the other end of the barrel of a tank or the seeker side of a Hellfire missile shot from a drone.

War and the Conversation between Moral Equals

In detailing a discussion of the Kosovo War after the fact with a Serbian friend, Michael Ignatieff laments, "We weren't even prepared to risk the life of our own soldiers in battle." His friend then says, "They were ready to risk the life of my wife and my children but not their own soldiers' lives." Ignatieff finishes the thought with, "If we had really fought them, face to face, he was implying, and if we had faced death, as they had done, then we might have had his respect."[34]

I recognize there are those who will express dismay over the idea that such respect from our adversaries is relevant in matters of war, but I think it most certainly is, for a number of reasons that have already been discussed. The idea of the innocence and equality of combatants in warfare forms the foundation of justified killing in war. But there is one reason having nothing to do with cerebral subjects of morality and ethics: the practical matter of someday finding peace. Koplow suggests this as one of the main reasons we exercise restraint in warfare, what he calls "self-deterrence." Why? For the purely practical matter of having to deal with former enemies when hostilities cease. History, he says, is full of bitter enemies not remaining so.[35]

Respect on the battlefield cannot be undervalued as a stepping-off point to solving the difficult issues surrounding a lasting peace. This is a cause for concern if we are indeed in a world of persistent conflict where we are simply setting conditions for desired outcomes. The first problem is misunderstanding the concept of persistent conflict and thinking there will never be a need for a settled peace. The second is how we can possibly set the conditions for desired outcomes in wars among the people if we discount the validity of maintaining respect among combatants on the battlefield. This is why the quote at the beginning of this chapter is so disconcerting. In no conceivable way is unmanned warfare speaking to those we are fighting in a way they understand.

Peter Singer recounts the story told by a Special Forces officer of a visit to the tribal regions of Pakistan where the U.S. launched just shy of two hundred drone strikes. The officer said, "One of the elders . . . went on to tell how the Americans had to be working with forces of 'evil,' because of the way that their enemies were being killed from afar, in a way that was almost inexplicable. 'They must have the power of the devil behind them.'"[36] Singer goes on to say, "While we use such adjectives as 'efficient' and 'costless' and 'cutting edge' to describe the Predator in our media, a vastly different story is being told in places like Lebanon, where the leading newspaper editor there called them 'cruel and cowardly' or in Pakistan, where 'drone' has become a colloquial word in Urdu and rock songs have lyrics that talk about America not fighting with honor."[37]

These illustrations do not serve to illuminate any grand concordance between our view of what we are doing with unmanned warfare and how it is received by those we are prosecuting with it, or even those on whose behalf we purport to use it. It seems, in fact, our views could not be farther apart.

We are faced with an interesting dilemma we have seen before and failed to recognize. Our failure in Mogadishu in the mid-1990s was perhaps a harbinger of things to come. We have allowed our concept of precision, an enabler of the trend allowing the application of force from ever-greater distances, and the disciplined way we go about our actions in war to obscure our view of how the enemy sees those same concepts and actions. We want to prosecute war humanely; it is why we place such a premium on avoiding collateral damage. Yet our enemies perceive this as a weakness and often use it against us. Coker says the U.S. military was demoralized in Mogadishu, "particularly as a result of finding that the other side would not allow it to be humane and even more that its technology did not always allow it to fight humanely."[38] Using civilians as shields and children as warriors complicates our Western ideas about how to fight, but the lesson is less about the tactics employed by Mogadishu gangs than it is about the mistaken idea that technology always raises the level of the discourse. It simply is not always true. Just as we could not use superior technology to fight humanely in Mogadishu, we cannot use our superior technology in Pakistan and Afghanistan to "talk to them in a way they can understand." The Western tradition of "instrumentalizing" warfare, in and of itself, is no guarantee of successful outcomes or even of setting the conditions for them based on mutual respect on the battlefield. Coker warns, "We will continue to instrumentalize war still further by diminishing the human factor. And as we continue down that path, we will find ourselves increasingly distant both emotionally and psychologically from other societies who have preserved the warrior tradition or find themselves more in tune with what Clausewitz called its 'true nature.'"[39] In addition, "the existential dimension of war was not only very real but vital in legitimizing it as a human activity. In instrumentalizing it, as Westerners have done today, we have made it purely utilitarian. That may be a virtue in our own eyes, but it devalues war in the eyes of others."[40]

If we are to fight wars among the people to set, in some way, the necessary conditions for whatever desired outcomes we may support, we have to consider how lethal unmanned systems support those interests. Fighting with impunity but "without honor" or in "cruel and cowardly" ways—or being backed by the "forces of evil"—is not likely to be successful in setting any conditions other than persistent conflict with each individual entity we engage and with no hope of a lasting peace. This is not to imply we cannot find effective ways of using advanced technology in these kinds of fights, only that what we have done so far does not seem to be having the desired effects. It is the classic case of "mirror-imaging"—seeing those you deal with only in terms fitting your own narrow viewpoint. Unfortunately, you cannot find the enemy by looking in the mirror—that is, unless your own worst enemy happens to be staring back at you. Krishnan warns us, "Robots can never 'win hearts and minds' and would likely indicate to the protected population that the intervening nation does not view the mission as very important, certainly not important enough to risk its own peoples' lives."[41] And so we return to the moral equality of combatants.

War among the people requires us to see through the eyes of our adversaries and think through their minds. If individual conflicts in a time of overall persistence are ever to be solved, if there is ever to be a lasting peace, mutual respect on the battlefield is where it inevitably must start. Recall our discussion of Rommel these nearly seventy years later as a man who fought well and whose primary Allied rival, Gen. George S. Patton Jr., certainly respected his skill and prowess in tank warfare. If we are to engage with warrior cultures, we ought to maintain a sense of the same, but failing that we must at least respect them for the way they view their lot and surroundings.

To the extent we can find ways to use emerging technologies to bolster this, we ought to do so. Where we find it adversely affecting the conversation as moral equals, we are bound to discard it. It would be the greatest folly to continue to procure military "capability" incapable of achieving the desired national objectives they were bought to secure. As Coker says, "It is only by relating to other people that we remain moral beings. If we choose not to relate we will no doubt act immorally whether we consciously elect to do so or not. . . . We are only moral beings in conversation with other people."[42]

The Enemy Within

If projections about the future nature of warfare are correct, our enemies of the future will indeed be those of the past. Even if war has changed, much remains the same. Gross says, "If war has changed, the underlying logic of settling political disputes by force of arms has not: war remains a costly last-resort means . . . when other means fail."[43] We will increasingly be faced with conflicts in which our adversaries maintain a sense of the warrior ethos and will, whether we like it or not, judge us for the way we engage with them in combat. They will judge us not on legal grounds, unless it benefits their cause, but on the ancient concepts of fighting with honor and courage. We might be in an age of persistent conflict as General Casey believes, one characterized by war among the people as General Smith believes, but certainly one where the likelihood of large force-on-force or even major combat operations such as those recently characterized by operations in Iraq and Afghanistan is very much diminished, as Secretary Gates has said. The question, other than how much stock we put in the predictions, is whether a move toward lethal autonomy supports the consequences of these characterizations of future war or not. If we maintain the vector and simply go farther along the path of massive and overwhelming force, it probably does not.

Asymmetry in warfare is a double-edged sword bringing, on the one hand, swift victory against like forces willing to engage on the same plane but without the technological mastery and skill in its use. On the other, it may provide no outlet for response in kind against adversaries too technologically deficient to mount a valid defense or counteroffense. The only ability of such forces may be to respond out of kind or in ways that inflict real damage but do not have the effect of decisiveness in battle. These strategies, such as using IEDs, tend to prolong engagement, wear down public support, and create mini arms races. An arms race of tactic versus tactic, as in the case of IEDs, may be illustrative of what could occur on an even larger scale.

Just as the use of UAVs by U.S. and other Western militaries has spawned competition and use of similar platforms by other states and organizations, a proliferation of robotic technology on the battlefield could spawn an arms race in such systems as prices, already lower than major weapon systems, inevitably come down with advances in technology. Our use of these technologies may then come back to haunt us as other entities gain the ability to use them in similar ways against us. Lacking that capability, our adversaries might seek other ways to respond.

They might eventually be able to target our unmanned systems operators, or they might lash out even further against our civilian population. We should not ignore the possibility that we are creating a situation that in some way legitimizes such actions by the simple fact we have denied any other response. This is the hypothesized consequence of the risk inversion created by increasing combatant impunity. Might a robot army, or an army with some number of lethal robots, exacerbate this possibility? Could we recognize it and live with its consequences if it did? These are questions we must answer before deploying more lethal unmanned systems into the fray. Those are future issues, probably near-future, but implications of unmanned warfare are being played out right now.

Speaking to our adversaries in a language of war they can understand is important. President Bush apparently thought so. Mutual respect for, or at least an understanding of, actions on the battlefield has important moral and practical implications. The concept of persistent conflict does not mean conflict with a single entity will last forever. It means there will always be the kinds of conflicts we see in the world today. Therefore, there is still a need to seek peace, to find a settlement to whatever grievances led to war in the first place. An ability to respect the way an adversary fights can be the basis of beginning the long conversation leading to lasting peace. Where we find disrespect for the ways of battle, we are likely to find long-lasting animosity that may erupt again in violence with little provocation. If we are to fight among the people for some time into the future, we need to figure out how to respect the warrior cultures of those we fight and show some sense of respect for the same in ourselves.

It is unclear how robotic technology may fit into this perceived paradigm of war. We are just on the verge of finding out all that it can and cannot do. Whether we are smart enough to recognize what it should and should not do is another question entirely. If we can hold on to the moral foundations of war and recognize the necessity for the moral equality of combatants, there is chance we will do the proper thing. If we fail to understand the moral foundation and disregard the practical implications, if we look into a mirror in search of the enemy, there is almost no chance of acting with prudence. Ironically it will be because we will have seen the enemy and failed to have recognized that he is—unfortunately and in spite of ourselves—the one staring back at us from the reflective glass.

8

AI, the Search for Relevance, and Robotic Jus in Bello

The trend is clear: Warfare will continue and autonomous robots will ultimately be deployed in its conduct.

—Ronald Arkin, director, Mobile Robot Laboratory, Georgia Institute of Technology[1]

Ultimately, we must ask if we are ready to leave life-or-death decisions to robots too dim to be called stupid.

—Noel Sharkey, professor of AI and robotics, University of Sheffield[2]

Depending on perspective, robotic warfare holds great promise or is a sign of the end of days. There is a middle ground too, with others thinking the only thing to do is continue the research and find out where it takes us. Some say the level of AI required to deploy lethal autonomous robots is unlikely or cannot be done. Others think there is no stopping the advance and eventually we will reach a "singularity" where humanity is transformed into its next incarnation.[3] That is the optimistic view. The more pessimistic view holds the singularity as the time humanity ceases to exist in its present form. J. Storrs Hall, a researcher in AI, is of the opinion we will never know until we try and that the only thing left to do is create the kind of AI the naysayers believe is impossible.[4] "Science," he says, "advances by experiment, not debate."[5] Indeed it does.

Debate is usually left until after the research occurs and experiments are complete, when we are left only with the realization of what we have done. This moment of discovery—sometimes horror—can find a no more telling example than in the aftermath of the first nuclear test. Manhattan

Project physicist J. Robert Oppenheimer later said he was reminded of the Bhagavad Gita as he watched the fireball roll upward from the New Mexico desert floor: "If the radiance of a thousand suns were to burst at once into the sky, that would be like the splendor of the Mighty One"[6] and "I am become death, the destroyer of worlds."[7] Whether we are on the path to such a humanity-altering event in the field of robotics is still unknown. Though I will not make predictions on when we may see lethal robotic systems with the *ability* to act autonomously, given the current state of research I think we are a long way off from having a fully capable system. That does not mean we will not field one earlier, as robotics researchers Patrick Lin, George Bekey, and Keith Abney raise in their paper *Robots in War: Issues of Risk and Ethics*.[8] I hope we will at least take time to have the debate prior to action. With that in mind, it is probably time to swing back around and talk once again about actual robots.

Science, Vishnu, and the Laws of War

Allow me to highlight some of the arguments for progressing toward robotics uses in warfare and briefly discuss some of the research and how it relates to *jus in bello* principles. At the same time, it will be necessary to talk a little about the science involved in the pursuit of some of the enabling technologies that might someday bring lethal autonomous robots to a battlefield near you—or perhaps if you are lucky, not very near at all. The intent here is not to write a text on AI or robotics engineering, as I am no expert in either field. What follows then, is really a narrowly focused, surface-level view of some of the research currently taking place. I will draw very tentative conclusions about current and near-future capabilities for how robots fit into two very important constructs of the laws of war, those of distinction and accountability.

Distinction, sometimes referred to as discrimination, flows from the humanitarian concept that we must avoid intentionally harming noncombatants. Noncombatant immunity stems from the idea that it is wrong to kill those who have not granted their consent to the possibility of death either by their taking part in hostilities or by their status as members of militaries of warring states or parties. This requires combatants to distinguish between themselves and noncombatants in order to meet *in bello* requirements for fighting justly. The criterion requires combatants to know that those they intend to kill are legal targets. It is "the central moral idea of just war . . . that only the combatants are legitimate objects of deliberate attack."[9] The requirement for distinction is codified in the 1899 and 1907 Hague Regulations, articles 22–28, and in the 1977 Geneva Protocol, article 1, "General Rule."[10] It loosely follows, due to the lagging nature of the law, which has not yet dealt with the status of systems like the ones we are discussing, that any lethal autonomous system on the battlefield would have to meet the standard of distinction. This concept is precisely the rationale behind the Ottawa Accord outlawing the use of antipersonnel landmines—inherently indiscriminate weapons—representing a significant precedent for other forms of lethal autonomous systems.

Accountability is the foundation of justice. Without an ability to hold people accountable for their actions, in war or in civil affairs, the rule of law breaks down. This makes accountability the cornerstone of just war theory and provides rigidity to the idea that war can be and is a rule-governed activity. As Michael Walzer says, "the assignment of responsibility is the critical test of the argument for justice. . . . If there are recognizable war crimes, there must be recognizable criminals."[11] The rule of law is built on the foundation that we can find and prosecute those we suspect for crimes against society and hold them accountable if they are found to have been responsible and can understand the difference between right and wrong.

It is the same in just war theory. Authorities are judged for their rationale in the resort to war, and soldiers are judged for their actions in the conduct of war. Heads of state have been convicted and punished for crimes against humanity, and military members have been convicted and punished for crimes of war. It is precisely the melding of responsibility and accountability that gives moral agency its weighty place in any moral choice but particularly in choices involving life and death. Moral agency—simplistically the ability to effectively weigh moral choices and make a decision based on a system of ethics—underpins the entire just war tradition. Once more from Walzer: "There can be no justice in war if there are not, ultimately, responsible men and women."[12] Accountability is then the second major wicket for the legality of lethal autonomous weapons.

For all the bluster of the "roadmaps" and the science fiction–sounding terms contained therein, the deployment of autonomous systems with the ability to purposefully take lives in battle basically boils down to meeting two criteria. Lethal autonomous killers will have to be able to distinguish combatants from noncombatants, and they will have to be part of an unbroken chain of accountability. In the first case we ought to expect them to do at least as well at distinction as their human counterparts—and probably better. There are those who believe we can design systems to achieve this goal. In the second case we will have to decide who the responsible party is, and it may or may not include the robot itself. As it turns out, accountability seems a relatively clear issue to handle, though we must first wade through the possibilities. The idea of robots meeting distinction is not quite as black-and-white. Perhaps a bit of background on AI will help illuminate the way in discussing the problem.

"Bang! The World Is Round"

Alan Turing, the father of AI, published an essay in 1948 titled "Cybernetics," which, for the first time, described AI in a recognizable way.[13] His ideas were formed while working on the top secret project now known as Enigma, the program designed to unravel Nazi cryptography and allow the Allies to decode communications among the German High Command.[14] Because he was employed in a classified enterprise, he lost out on bringing his concepts for early computing into the public eye, but he remains relevant today for the test that may someday herald the arrival of human-like intelligence in a machine. The "Turing Test," as it is now known, is really a question that asks, "Can a computer fool an interrogator into thinking it is human during an interview across a teletype link?"[15] Dr. Hall explains further: "The genius of the test is that . . . the use of language in an interactive setting means that any suspicion of weakness entertained by the judge can be followed up and tested to any desired depth. There are some skills that cannot be conclusively demonstrated over a teletype, but it should be pretty hard to convince someone that you're intelligent if you're not."[16] Now there is no need for human-like AI, what I refer to as "strong AI," in order for machines to meet distinction in combat. However, there is something to the Turing Test that is valid for all forms of AI and will certainly be required for the lethal autonomous machines of the future battlefield, if they are ever deployed.

The *real* genius of the test is a bit subtler than Hall explains, though Turing stated it explicitly. What he learned from his work on Enigma was that "intellectual activity consists mainly of various kinds of search."[17] What he does not state explicitly, but what every elementary school student who has ever used the Internet to research a report implicitly understands, is that the intellectual activity consisting of different kinds of search ultimately comes down to sorting through data for relevance. Intelligence cannot be proven simply by assimilating irrelevant data, just as sanity cannot be proven only by reciting objective truth. AI researcher Clark Glymour explains Søren Kierkegaard's "objective madman" as a man walking around with two glass balls on his coattails who, every time they collide, exclaims, "Bang! The world is round." The objective madman speaks an objective truth, but "he fails to draw the relevant conclusions."[18] Though he repetitively speaks the truth, he is judged insane for his continuous claim, much as we might judge a computer that cannot sort relevant data to be less than intelligent. The issue of relevancy—of turning masses of data into useful information by sorting through all those things that are irrelevant—sets up one of the major difficulties in programming AI. It is called "the frame problem."

Researchers John McCarthy and Patrick Hayes first described the frame problem in a 1969 paper titled *Some Philosophical Problems from the Standpoint of Artificial Intelligence.*[19] Stated simplistically, the frame problem answers the question "Given a lot of information, some of which may need to be used to do a task, what is the relevant information for the task?"[20] Hayes describes it more completely by first considering a simple axiom about basic movement:

> IF the agent is in room *r* in state *s*,
> AND IF the door *d* from *r* to *r1* is open in state *s*,
> THEN the agent is in *r1* in state *gothru(d,s)*[21]

The axiom describes the act of going through a door into another room. Easy, right? On the surface it seems so, yet think of all the things that might have changed but did not. The immediate consequences are clearly inferred, "but what about the immediate non-consequences?" He explains: "When I go through a door, my position changes. But the color of my hair, and the positions of the cars in the streets, and the place my granny is sitting, don't change. In fact, most of the world carries on in just the same way that it did before, or would have done if I hadn't gone into room *r1*."[22]

The complicating factor is that most of the things that did not change could have. They are "relative to the time instant, so we cannot directly infer, as a matter of logic, that they are true in state *gothru(d,s)* just because they were in state *s*."[23] It seems there ought to be an efficient way to say what changes when an action takes place without listing all the things that do not change, yet there does not seem to be any other way. Hayes says, "That is the frame problem."[24]

Cognitive scientist Daniel Dennett describes it in a now famous way, and it bears repeating here. He dreams up a robot named R1 that (or is it "who"?) is taught that its spare battery is in a room with a time bomb that is about to go off. There is a wagon in the room, and the battery is on the wagon. R1 quickly decides to pull the wagon out of the room and does so. Unfortunately, the time bomb was also on the wagon. Ironically R1 knew the bomb was on the wagon, but it could not reason that pulling the wagon out of the room would bring both the battery and bomb, because it could only consider its actions in relation to the battery. The programmers realized that they must program an ability to consider the side effects of actions, so a new version of the robot, R1D1, begins to consider all the things that will and will not change. As it finishes figuring out innocuous items such as how the color of the room will stay the same and the number of revolutions the wagon's wheels will make, the bomb goes off. Finally the programmers understand they must teach the robot the difference between relevant and irrelevant effects. This time a third version, R2D1, stands there seemingly catatonic. When asked why it appears to be doing nothing, it responds that it is busily making lists of thousands of irrelevant things. The bomb goes off.[25] Apparently R2D2's programmers finally solved the frame problem, or they touched it up with a bit of Industrial Light and Magic.[26]

The frame problem may be a significant hurdle to clear before lethal autonomous robots are set about the battlefield. Of course, there is always the chance of some major breakthrough we are not anticipating. It must also be said not all AI researchers consider the frame problem to be a major impediment. Some claim it has never been an issue.[27] Patrick Hayes says that is because most AI research deals with problems that never encounter the frame problem; they are restricted to very limited operation in an environment that does not change. The problem arises only when programmers are thinking about a changing, dynamic world.[28] Well, the world of war is nothing if not changing and dynamic.

Air combat, as an example, is a dynamic, three-dimensional, multivariable, nonrepeatable death match. Adversaries may come from known or unknown locations. They may come in larger or smaller numbers than you expect. The former is difficult, but the latter keeps you wondering where the rest are. They may carry ordnance you did not expect or execute tactics you are unfamiliar with. Then, because they are human, they make mistakes or have flashes of brilliance making every air-to-air engagement, in the real world and in training, like the geometrical complexity of snowflakes—there are never two exactly alike. War is an environment that never stays the same. In a simple sense, there are ROE that must be followed, and this provides warfare with some structure.

However, it is not always simple. Though they are called "rules" of engagement,[29] ROE really form a constraints-based system. There are certainly if-then statements to be settled. This is the domain of rules-based systems. For example, an aircraft is or is not replying with the correct identification code. Once this is known, there are certain follow-on actions. But it is the constraints that either allow or prohibit the use of lethal force. As an example, an aircraft may meet all the rules to classify it as an enemy but may not be engaged for some other reason; for instance, it has not committed the hostile act required by the ROE or is flying a profile that is indicative of a defector. It is bad form to shoot down pilots who would rather join your side. Constraints-based systems put limits on lethal force that without the constraint, in this case written in a portion of a legal document called ROE, would otherwise be legal and justified acts of war. Add ambiguity to the constraints of ROE, and you are beginning to get a feel for the complexity of the use of lethal force in combat.

Most probably assume ROE are unambiguous. Most combatants wish they were. Unfortunately ambiguities abound. There are ambiguities in what identification systems will tell you. There are ambiguities about where wreckage might fall or sink and whether where it does is considered the right place, usually on the correct side of some border. There are ambiguities about where all the friendlies are and whether you might be shooting lethal ordnance through their formations. In air combat this last one is usually cast away in a giant roll of the dice aviators call "the big sky theory": big sky, little missile. In trying to quantify the big sky theory, we talk about a "clear avenue of fire" as a limitation to shooting. How is it satisfied? By saying to the shooter, you must ensure a clear avenue of fire. I am not sure it could be more ambiguous if we tried to make it so.

There are also times when even though all ROE are met, it just is not a good idea to shoot. Something makes the hair on the back of the neck stand up and tells you that a piece of the puzzle is missing. Then there are times when even though all ROE are not met, the pilot may decide to fire anyway. Those are decisions made on what we call "situational awareness"—the ability to search the data, turn it into information about the current environment, and enable a well-reasoned decision, in this case one that involves taking lives. There have been times when I met all ROE and decided not to fire. I know those who did not have important pieces of that matrix and chose to engage anyway. Those choices, to fire and not to fire, all turned out to be correct. In the end, all of these situations rely on the human mind's amazing ability, with very little power input, to sort the relevant from the irrelevant "in various kinds of search." It is hard to imagine a robot with the ability to sift through the relevant and irrelevant issues involved in making life-and-death decisions, but that does not mean it cannot be done. As Hall so aptly states, science moves forward through experiment, so enough of debates for now.

"War in a Can": Distinction and Robot Experiments

There are clearly good reasons to explore the possibility of lethal robots. Ronald Arkin is a leading researcher on ethical lethal autonomy. His research in this area was funded by the U.S. Army as part of a three-year program that asks, in part, "Should soldiers be robots? Isn't that largely what they are trained to be? Should robots be soldiers? Could they be more humane than humans?"[30] The impetus for his research is in defining constraints for ethical lethal behavior in robots. Its rationale was to address the advantages robots might bring to the fight. Among these are their ability to act conservatively, to carry better sensors, and to avoid "scenario fulfillment" (a phenomenon that sometimes causes humans to ignore contradictory information about a current crisis/scenario). Additionally they lack emotion that could cloud responses, are faster processors with broader integration of information, and have a capability to monitor behavior and report when breaches occur.[31] Arkin is quick to point out he is only accomplishing research; it is pure science progression. He has no strong opinions on the "should" questions of lethal autonomy, and his only interest in such arguments involves the development of a robust background knowledge. He does not want to see lethal robots deployed without the research to back up how they might be controlled and used.[32]

Other advantages, at least in the short run, include things we have already discussed, such as keeping soldiers out of harm's way. As Steve Featherstone says of robots, "when they are destroyed, there are no death benefits to disburse. Shipping them off to hostile lands doesn't require the expenditure of political capital either. There will be no grieving robot mothers pitching camp outside the president's ranch gates. Robots are, quite literally, an off-the-shelf war-fighting capability—war in a can."[33]

So what might we get for our "war in a can"? Arkin's research produced robot programming that constrained lethality under ethical judgments in three basic forms. Actions were in some combination of obligated, permissible, and forbidden.[34] (It was all computer-based; no robots or humans were harmed in the making of his book.) The concept of a constraints-based system is the same here as was discussed above. In situations where firing was forbidden, the robot would not act. Though Arkin discusses an option to allow for the supervisor to override the system in this case, he does so only with reservation.[35] His basic premise is that robots may perform more ethically than humans, and in allowing an override of this nature he may be limiting one of the main advantages of lethal robots. In instances where firing is permissible, the system must also seek obligation to do so. Permissibility is not enough.[36] Consider the defector scenario above, where the object is identified as an enemy but firing is constrained by a lack of obligation to do so. He says "obligating constraints provide the sole justification for the use of lethal force within the ethical autonomous agent. Forbidding constraints prevent inappropriate use."[37] Arkin used unclassified versions of recent ROE to set up a constraints-based ethical system, then programmed and tested it. He concludes it is possible to program some kind of ethical standard—a conscience, if you will—for machines to follow in lethal actions.

Importantly, Arkin's concept of robot conscience is really a metaphor. It represents a means for ensuring the robot operates within the rules and constraints of society.[38] He rejects onboard rules derivation, noting the dangers of letting human soldiers do their own ethics derivation on the battlefield.[39] Apparently the robots will not be reading Kant either. He also notes a robot cannot be a moral agent because moral agency requires accountability.[40] This is a particularly salient point, not only for what it says of accountability but for what we have discussed in previous chapters about the moral equality of combatants and the moral nature of war.

We now turn to military research laboratory experiments in autonomy from the Office of Naval Research. The ONR sees operational advantages in autonomy, though it is not investigating lethality as far as this unclassified research shows. The ONR seeks to provide new capabilities that expand the operational envelope of naval forces, provide force multiplication or replace existing capability with less expensive alternatives, and provide for survivability of systems through distributed assets and redundancy. They are also looking to reduce the need to place personnel and high-value assets in high-threat areas, and reduce manning and communications requirements. These are dubious claims given the tail-to-tooth ratio of unmanned systems and the worldwide trend toward more bandwidth. The ONR is also attempting to reduce the cognitive load on the war fighter and increase the quality and speed of decisions.[41] An ONR engineer briefed three projects to the Robotics and Autonomous Systems Industry Study at ICAF in March 2011. These projects were in the fields of intelligence, surveillance, and reconnaissance or in search experiments integrating a human supervisor managing up to eight unmanned vehicles of different types.

The experiments all presented some form of supervised autonomous level of control. The supervised autonomy missions exhibited autonomous collaborative behaviors such as team task allocation and scheduling, secondary task optimization, and distributed collaborations under limited communication environments.[42] The operator's tasks were often complicated by incomplete intelligence. This lack of perfect intelligence information—very realistic—often forced on-the-fly tactical reprioritization for the operator and autonomous task allocation for the vehicles.

In one such experiment, a virtual UAV and UUV found and tracked an actual surface vessel of interest leaving a port and were able to autonomously hand off the tracking to a USV, which was then able to identify the vessel and continue shadowing it. In another experiment, four UAVs searched a wide area, collaborating on completed and incomplete tasks while constantly retasking individual UAVs in a shared effort even under limited communications. In the third experiment, a multitude of small UGVs, along with a few larger Segway-platform UGVs and a UAV, tracked a human test subject wearing an orange vest around an urban environment via onboard sensors. (The unmanned vehicles used the vest as the identification tool; combat-quality identification was not the subject of this test.) The goal was for the robots to autonomously establish a communication network, find the subject, and maintain contact by sharing information.[43]

These are impressive demonstrations of autonomous robots performing valid military and security functions. However, some of their more humorous moments serve as a "reality check" on near-future capabilities. In one instance, a Segway robot was unable to recognize one of the smaller robots, classifying it instead as a frangible obstruction it could drive over.[44] This might have led to one of the first instances of robot fratricide had the human experimenters not intervened. These kinds of incidents in robot experiments invariably lead to concerns about the plausibility of the systems and the AI that must drive them. It is the kind of thing J. Storrs Hall was reacting to with his comment about the advancement of science. Still the concerns are worth noting.

Armin Krishnan, in *Killer Robots*, says there are three main deficiencies of autonomous robots with respect to the requirement for distinction in lethal systems. Machine perception—the Segway driving over a smaller robot—and control software that would have to be too complex and would be "brittle and more likely to contain bugs"[45] are two of the issues. He also thinks it unlikely AI can make robots instinctively find relevant elements and implications of a situation—the frame problem. "As a result," Krishnan says, "military robots would either be too slow to make them militarily useful or would be prone to use force indiscriminately and disproportionately, as they would often miss important details or incorrectly interpret situations."[46] Krishnan shares Arkin's concern that they may be deployed too soon, saying technological development of robotic weapons may proceed faster than researchers can make them sufficiently safe before battle.[47] Their concerns are valid.

The DOD has a pretty good record of sending new toys into battle while still in the test stages. It was true in the Gulf War of the E-8 Joint Surveillance Target Attack Radar System aircraft, both the Predator and Global Hawk UAVs saw action before testing was complete, and none of the ground robots in CENTCOM were fully tested programs of record until the M-160 showed up. None of these save the Predator were armed—and that capability was tested before combat—but whether it is likely to remain this way in our time of persistent conflict is certainly worth debate.

At this point it is probably too soon to make predictions on how or whether difficult issues in robot intelligence will be solved to allow for meeting the requirement of distinction. Early strides are being made in research laboratories at universities, in the private sector, and at military research facilities. It might be possible to design systems that can act as if they follow some ethical code, although the terminology of robotic ethics is probably well before its time. Machines acting inside constraints programmed by humans are not acting ethically in any real sense of the word. This, like Arkin's idea of robot conscience, is currently metaphorical. Autonomous, collaborative nonlethal behaviors are being tested, and at least one ONR researcher believes one such system could be fielded now.[48] In no way do I mean to diminish the amazing strides in robotic technology or turn a blind eye to its possible benefits, but impressive as they are, many of these experiments demonstrate what Alan Turing understood more than sixty years ago.

The search for relevance is a key component in intelligence. Turning R2D1, whose battery is blown to bits while it endlessly categorizes irrelevant information, into R2D2 is not an easy task. For now humans have a monopoly on intelligence of the sort required for satisfying distinction. As Krishnan says, "Humans are better at discriminating targets not because their vision is better, but because they understand what a target is and when and why to target something or somebody."[49] We are probably some time away from "Bang! The world is round" to "Bang! You're dead" at the hands of an autonomous robot, but we might get there. For now, let us turn our attention to the clearer problem of accountability.

Benign Psychopaths: Accountable Robots?

Ask any six-year-old in the immediate aftermath of some catastrophe involving spillage, breakage, or a sobbing sibling, and the response will almost certainly be some form of "It wasn't my fault." The human propensity to shift blame is seemingly learned very early on. The reason for this is fascinating. It is what underpins the basic rule of law holding society in a tenuous suspension and preventing the descent into total anarchy and chaos. The reaction of a six-year-old to shift blame seems to be driven by his knowledge of the consequences he will face when eventually "found guilty" of the heinous crime of knocking over his milk due to lack of attention to the task at hand. It is this knowledge of consequences and the placement of value on them, or rather the relative value of avoiding negative outcomes, that creates this suspension. What holds us together as a society bound by the rule of law comes down to the idea that we all buy into the concept. If it were not so, it would not stand. This is probably why psychopaths, who cannot empathize on a personal level and do not base actions on consequences, are so unnerving to us. They show us the possibilities of an existence we find abhorrent and terrifying. It is why, for the greater good, we all submit to the judgment of our peers. We accept we will be held accountable for our actions, even if our first reaction may be to deny fault.

Questions of accountability for robot "warriors" most often come down to whether or not we can hold a robot accountable, and if we cannot, then who should stand in. It is interesting, but it is really not the most interesting question or even the most difficult to answer. There is currently no moral or practical basis by which our machines can be held accountable, but it does not stop the hand-wringing over the possibility.

One of the most chilling lines ever recorded on film is "I'm sorry Dave. I'm afraid I can't do that." HAL, the spaceship's computer, has just decided not to let astronaut Dave Bowman back aboard in Stanley Kubrick's classic 1968 science fiction film *2001: A Space Odyssey*.[50] He, or it, had just made a lethal decision, significant for the fictional life of Dave Bowman but even more so for the collective conscience of humanity. Over forty years after the film's release, HAL is still a metaphor for the unexpected, perhaps detrimental nature of the progression of technology. The question we are about to face is best posed by Daniel Dennett, creator of the poor robot who cannot retrieve its spare battery, in the title of an essay called "When HAL Kills, Who's to Blame?"[51]

Given that our current state of AI has so far produced "robots too dim to be called stupid,"[52] it seems almost ludicrous to point out we are nowhere near having robots for which accountability for actions is even a relevant concept. However, history is full of predictions and judgments about the future of technology that now seem ridiculous in retrospect. It is not my intent here to say the argument has reached its ultimate end state. I will highlight a differing view, again from J. Storrs Hall, that the vector of technology is taking us toward conscious AI that can understand consequence and therefore toward accountability. For now, with robots who cannot recognize the difference between another robot and a frangible obstacle, I will simply say it is not feasible nor in our interest to push accountability onto our machines. They currently do not have the necessary frame of reference to understand or make decisions based on consequences. The robots we have today are—so far—benign psychopaths.

In the essay mentioned above, Dennett says accountability is not something we can simply grant. There is a foundation, again starting in the realm of morality and nicely paralleling just war and civil law concepts as you might expect. There was a subtle point in an earlier example where I mentioned we hold people accountable under the law if they are capable of understanding the consequences of their actions. This drives us to hold hearings on mental competence, for instance, in cases where learning-disabled individuals or those born with certain genetic conditions affecting mental capacity are suspected of committing crimes. In short, we require intentionality and consequence recognition before punishing for actions outside societal norms. This is why collateral damage in war is justifiable under the doctrine of double effect; the damage is foreseen but not intended. Dennett says, "Higher-order intentionality is a necessary precondition for moral responsibility . . . [meaning it must be] capable of framing beliefs about its own beliefs, desires about its desires, beliefs about its fears, about its thoughts, about its hopes, and so on."[53] Our machines are a long way from this ability. In his research, Arkin requires an accountability link to humans by virtue of a check box in the operator's program interface authorizing lethal force in a given instance and under given constraints.[54] Robots are not, he adamantly points out, moral agents.[55] Without moral responsibility we should not—in fact cannot—hold them accountable.

An equally valid view of the inability to hold robots accountable is that of precedent. They are machines put to use by humans and for humans. They are programmed by humans to do certain things under certain normally static conditions where the results are known (so as to avoid the dynamic conditions that invoke the frame problem). We have never given responsibility and accountability to our machines, so why start now? Steve Featherstone writes in "The Coming Robot Army," "Robots invite no special consideration under the laws of armed conflict, which place the burden of responsibility on humans, not weapons systems. When a laser-guided bomb kills civilians, responsibility falls on everyone involved in the kill chain, from the pilot who dropped the bomb to the commander who ordered the strike. Robots will be treated no differently."[56] I think for the near future, perhaps out to the limits of the DOD's robotic "roadmaps," he is right, but this is a contentious issue.

Higher-level intentionality means there must be some form of free will. We currently have no machine with this capability, but computer scientists are thinking about the possibility. The difficult part of programming free will is that it must deal with a philosophical conundrum about the nature of the physical world and ideas about choice. First, and inherent in Yale professor and computer scientist Drew McDermott's free will architecture, is the idea of the universe as deterministic—that is, it seems to follow physical laws or is "probabilistically random, which is also in accordance with physical law (e.g., chaos theory or quantum mechanics)."[57] Second is that we have a strong idea that we have the ability to make choices.[58] These seem to be contradictory and remind us of Orwell's statement that our biggest mistake is to assume we are autonomous beings.

The problem with programming free will is that the robot needs a deterministic model of the world as a frame of reference but must see itself as nondeterministic—capable of choice—while "living" inside the deterministic world.[59] This is exceedingly hard to do. It turns out that McDermott's model—the world as deterministic but the creature exhibiting choice—forms a theory that intelligent beings "come with a built-in freedom/determinism problem" as long as they are "sufficiently introspective."[60] If a robot could be built along these lines it would have the "seemingly contradictory intuitions of determinism and free will."[61] It would share the Orwellian belief we all have that it in fact does have free will, and it would then "act in the same sense humans do."[62]

Would we then have an accountable robot? The clear answer is "yes." Hall goes so far as to say such robots would then be able to consider the negative outcomes and would alter their future behavior. He says, "So, from a purely practical standpoint, it makes sense to punish robots with McDermott's architecture. This seems to me to be a clear indication that the theory is on the right track."[63] We are not there yet—nowhere close—but it might be out there in our shared future. Drawing us back from the future and into the present, we have to ask who we will hold accountable if the things doing the killing will not yet be.

Robert Sparrow, professor of philosophy and bioethics at Monash University in Australia, says there are three possibilities for accountability for robot actions: the manufacturer/designer, commander, or the robot itself.[64] I have dispensed with the latter, though again this is not the end of the argument by any means. It is for others to take up among the inevitable advancements in AI. Returning to Sparrow, I will elaborate on the first category.

The manufacturer should be judged separately from the designer, the programmer, or the engineer responsible for designing the part or code that somehow failed to perform as expected and is now forcing us to consider accountability at all. There is a chance the designer knew of some fault or design flaw and followed an appropriate procedure to report it. The manufacturer may choose to ignore the warnings, produce the robot, and fail to warn the end user of the issue. Perhaps the most infamous incident of this kind resulted in the destruction of the space shuttle *Challenger* in 1986. Morton Thiokol engineers warned of a faulty O-ring design on the shuttle's solid rocket boosters but were ignored by a management bending to pressure from NASA for no further launch delays.[65] The seals failed and allowed hot gas to ignite the shuttle's liquid-fuel tank, which then exploded. It seems clear we have to separate the manufacturer from the designers of the system, but even then it may not be a good idea to make them responsible for the later actions of lethal robots.

There is no precedent, as I have said before, of holding machines accountable for actions. There is also no precedent for holding the makers of military machinery responsible for the *normal* use of their equipment in combat. We hold them accountable for knowingly providing faulty equipment or for not providing information necessary for proper risk assessment, as in the case of the *Challenger* disaster. However, if their equipment performs as designed, manufacturers are not responsible for questionable use or the errors of their customers. The reality is that weapons malfunction; it is an imperfect world. Their human operators and authorizers of lethal force are aware of the possibility and accept both the risk and responsibility for things that go wrong in such weapons. The makers are not accountable for the unintended destruction. Though the company may be asked to redesign some portion of the system to make it more robust, manufacturers, or even the designers, are not criminally negligent, nor are they guilty of war crimes if the weapon falls into a wedding and kills innocent civilians. If the manufacturer gave appropriate information regarding the risk of using the system to the military, how can we hold the manufacturer responsible?[66] It gets even trickier with the advent of autonomy.

In simple cases of autonomy (if there is any such thing), where systems are given specific parameters or constraints to operate under, programmers and designers often know how the systems will react.[67] Science and engineering are, after all, based on predicting repeatable results. But as the systems get smarter, as AI progresses, the concept of programmer accountability becomes more tenuous. If the programmer gave the system appropriate information but the autonomous system is supposed to make its own decisions, it seems hard to argue how the programmer could be responsible for the actions of the machine.[68] The actions of a free-will robot such as the one Hall proposes based on McDermott's model could not possibly be blamed on the programmer, unless we decide as a society that the creation of such intelligence is, in and of itself, criminal negligence or reckless endangerment.

Holding manufacturers and designers responsible for the later actions of lethal robots is logically unsatisfying, but it is most likely also practically prohibitive. Discounting for now the massive change in the laws of war required to hold corporations or individual employees accountable for the actions of their weapons during acts of war, consider the market forces driving corporate decisions. It is hard to imagine a company willingly entering a market providing weaponry to militaries if they knew they stood the chance of being hauled to The Hague and prosecuted for war crimes for events that occurred after their products left their control. The manufacturers of the Patriot missile system were not prosecuted for the actions of the system in mistakenly identifying two friendly aircraft during Operation Iraqi Freedom, as discussed previously, even though the misidentification led to fratricide. This is probably the way we should expect it to be. Holding manufacturers or designers accountable does not seem to be constructive.

This leaves only one option for accountability. I can see no supportable case, in military operations, where the commander or the authority that authorizes lethal force will not be accountable. But this leads to the most interesting issue and one that has not been dealt with sufficiently. How do we propose to grant the responsibility that ought to accompany accountability to that commander authorizing robotic lethal force? We move now to the issue of military accountability for lethal autonomous systems.

Fires from the Sky and Military Accountability

When I deployed to Operation Iraqi Freedom, I gave a "going to war" speech like I suppose most commanders do. I'm sure it was nothing as colorful as Patton or Robin Olds would have done, but a key piece of it was about trust and accountability. I trusted my pilots because I knew they had been trained well. I hoped they trusted me to make the right decisions and do things right. A part of doing things right included my commitment to back up their valid decisions, as long as they were well reasoned and within the ROE, but another part was my commitment to act swiftly and ruthlessly against anyone who broke trust and discipline. Mistakes are costly in combat. Willful actions in defiance of known standards and rules can be devastating, particularly in wars among the people. Errant tactical decisions often have strategic effects. They understood I gave them both responsibility and accountability. The two together are the least we ought to accept.

Military commanders are certainly accountable for their actions in combat and for the actions of their men and women. There is a very good reason for this that most commanders accept as a matter of fact: it is part of the contract of command. They are accountable because they are also responsible. They are responsible for the training and equipping of their unit. They are responsible for selecting the leadership positions below them. They are responsible for grooming their chain of command to act consistently in their absence with what they would have done had they been there. A unit is a reflection of its commander. He gets the blame for their errors, and he gets the accolades—which good commanders always share with those who have done so well for them—when they succeed. Accountability, it turns out, is a two-way street, but the complexity of autonomous systems may strike this contract at its core. As the systems get more complicated, responsibility gets more diffused.

In the complexity of air combat detailed above, numerous systems feed information to the pilot. She has to understand the system, its intricacies and limitations, in order to draw valid conclusions. For current systems, such as identification tools in modern fighter aircraft, our sense is we can teach the pilot the things she needs to know to still make the proper decisions and maintain her accountability. If she shoots down the wrong aircraft, perhaps one carrying civilians, because of a failure to understand what the systems might have been telling her, she is still culpable. How long this may remain the case is uncertain, particularly as autonomy moves farther along the spectrum toward supervised or full autonomy. Steve Featherstone says, "It will become vastly more difficult, however, to assign responsibility for noncombatant deaths caused by mechanical or programming failures as robots are granted greater degrees of autonomy."[69] As the complexity of the system increases with an increasing level of autonomy, there may come a point where military commanders cannot maintain the level of responsibility required to keep continuity in the accountability chain.

Currently the authority to employ and accountability for the consequences eventually reside in a single person. Depending on the type of operation, that single person may be at different levels in the command structure. For a platoon out on patrol, it could be the platoon leader. For the operation of taking out a key Baathist communications node in an urban area, the authority may reside at the air operations center director level. In an example of a current scenario, and one that may answer one of the legal questions about unmanned warfare raised previously, the ground commander in charge of a defined battlespace owns all fires in his area of operations (AO). This includes rifle fire from individual soldiers, indirect artillery fire, and weapons from airborne platforms. One Special Forces lieutenant colonel said he makes no distinction between artillery, bombs from a fighter or bomber, or missiles fired from a drone. He owns them all, and nothing gets shot in his AO without his approval.[70] It could be bombs from an F-15E or rocks from a slingshot. He is responsible and accountable. This is a powerful legal argument for armed unmanned vehicles. If the ground commander is responsible and it truly does not matter where the ordnance comes from, there is a defense of the legality of remote fires such as those accomplished today by remotely piloted aircraft.[71]

Ground commanders are willing to take on this role because there is another kind of contract in place. They respect and trust the training and discipline of their joint partners just as they trust in the actions of the team they have trained. Autonomy might change that dynamic by removing the implied responsibility for robot actions.

There are analogies today where there is no implied responsibility, even for commanders. A fighter squadron commander is not responsible for the loss of an aircraft due to mechanical or engineering deficiencies (unless he is also responsible for the maintenance that caused the mechanical failure). He has no control over the engineering and procurement process that delivered the aircraft. He is, most likely, not an expert in the manufacturing or design of modern fighters. He has to trust someone else to do the proper testing and evaluation. If something beyond his control goes wrong while he or one of his pilots is flying, he does not have to answer for it. However, he is responsible if his pilot was attempting tasks he was not qualified for or was incorrectly trained to perform. Autonomy takes this one step farther.

Sparrow has said it would be immoral to field autonomous robots because the manufacturer, designer, or user will all lack accountability.[72] Lin, Bekey, and Abney say, in characterizing his argument, "No one can reasonably be said to give morally responsible consent to the action an autonomous robot performs; so no one is responsible for the risk such autonomous robots pose, and thus it is immoral to use them." They deconstruct the fielding argument by invoking a kind of morality for the robot called "slave morality," where "the robot cannot be blamed, for it really is 'merely following orders,' subject to the limitations of its programming. It could not become a *morally* autonomous 'law unto itself' and serve its own ends; hence, it cannot be held morally responsible for its actions."[73] The responsibility in this case falls back to the user. While the argument is effective in questioning the morality of fielding these systems, it does not directly address the more subtle point of Sparrow's argument. At its heart lies the diffusion of responsibility that occurs with increasing autonomy.

It is unlikely the commander of autonomous lethal systems will have anything to do with programming them, unless we get to the point of having the lab-coat warriors discussed previously. There is no analogy to the current system where the commander is both responsible and accountable due to his involvement in the training of his soldiers. Robotic warfare may be ready-made "war in a can." It is also unlikely he has the ability to understand all the ramifications and consequences of the advanced programming for all the systems he may have under his control. This inability obviously becomes more pronounced as the systems gain greater degrees of autonomy. At this point, holding the military commander responsible is "nonsatisfying" as the systems have the capability to make decisions.[74] Responsibility may eventually become so diffused it can no longer serve as the basis for accountability. Then there will be no one able to step up and say, "This is my doing; I am responsible."

There may be a time when no person or thing is accountable for the actions of our lethal machines. At some point, AI may advance to the point it exhibits a form of free will, and if this occurs commanders will prudently insist on the same level of training needed for human war fighters. Any savings in money and time because of the promise of repeatable results—war in a can—for autonomous robots will be gone. Maybe then we will readdress the accountability of machines, but they will not really be machines as we know them today. They may have achieved that higher-level intentionality Dennett speaks of. We will have to deal with the deep philosophical questions of what it means to be conscious. Ironically, at that point we will have conceivably "progressed" right back to where we started. We, or our robots, may once again inhabit the world of battlefield fallibility due to intentionality, moral dilemma, and the judgments they produce.

The Infinite Regression

If we are to employ lethal autonomous robots on the battlefield, we will need to ensure we have answers to two important aspects of the laws of war. Our robots will be required to distinguish between combatants and noncombatants, and we will have to define the chain of accountability for our autonomous killers. The first is laid down in legal precedent respecting the immunity of noncombatants and in outlawing indiscriminate weapons in the past. Distinction, so far imperfect when left to humans, may actually be more difficult for machines, at least in the short run.

AI programmers are faced with numerous challenges in designing systems with an ability to handle the dynamics of war. The first of these is the frame problem, which stems from the requirement of any intelligence worthy of the name to be able to sort through all available data in search of the relevant information. Without such a capability, robots may be unable to make the appropriate decisions for any given scenario. The frame problem comes up in dynamic situations where change is always a possibility, though not necessarily present. These situations complicate AI programming because of the enormous number of possible combinations of changes that may or may not be relevant. One researcher claims to have successfully tested ethical lethal autonomous systems. These are rudimentary experiments that are not operationally deployable right now, though the research holds some promise. It is important to understand that these robots are not really acting with ethics or any so-called conscience. These are metaphorical terms. They denote only an ability to act in accordance with programmed norms, which are in turn devised from given rules and constraints.

Another issue with robotic distinction is making sense of sensor information so as to avoid running over other robots, for instance, or shooting the wrong thing. The programming code required to make all of this work has to be almost perfect, yet it will likely be fragile and could affect robot function. A few AI researchers are looking beyond solving the frame problem to a time when AI progresses to the point it can exhibit free will. It is probably a long way off, and recent experiments in autonomy, though successful in moving the science incrementally forward, are not yet to the point where lethal robots meet the distinction criterion. This gives us time to work out issues of accountability.

Accountability is the basis of the rule of law. In order to deploy a lethal robot in good conscience, if that is possible, we have to come to some conclusion on whether a robot can be accountable for its actions, and if it cannot, who can be. There is no precedent for machine accountability, and that might be enough in ruling out robots. There is more, however. Currently our robots lack higher-order intentionality and moral agency. Both are preconditions for moral responsibility, so it seems unlikely, at least at the current state of AI, for robots to be accountable. This is not the end state of this argument. It will have to be revisited with advances in AI, but for now, after disqualifying the robot, three choices remain for who can take up the mantle of accountability.

The manufacturers, the designers, or the military commander are the remaining options for accountability on the battlefield. Manufacturers are an unlikely choice given the precedent of current operations. As long as the manufacturer provides all necessary information to calculate risk and does not knowingly provide deficient products, the military has always accepted responsibility and accountability for the actions and uses of its weapons of war. There is no compelling rationale to alter this arrangement, just as there is no compelling argument to push accountability onto the designers. While it is important to keep these two under separate consideration, primarily for negligence or deficiency in one not reported on by the other, it would be hard to make a case for an accountable programmer given that the robot operates within the parameters of the programming. It gets slippery with increasing autonomy since the machines are deciding on certain actions. This further removes the programmer from the accountability chain. It is another aspect of the insidious nature of increasing autonomy, which tends to diffuse responsibility and may ultimately affect accountability.

With the military commander as the only remaining option, the issue of responsibility comes to the fore. A commander is accountable because he is also responsible. It is his job to prepare a unit for combat and to understand the technology he plans to employ with a great enough degree of expertise to understand its strengths and limitations. He cannot make the proper risk-reward tradeoffs without that knowledge. However, the complexity of lethal autonomous systems may eventually sever the required responsibility. Will a commander understand all the autonomous systems in his charge with the degree of expertise required? He is unlikely to have been a part of the programming team, and increasing autonomy may reduce his ability to achieve repeatable results. If this occurs, he would be prudent to spend the time and money to train his army of robots, particularly if they have some form of learning autonomy, just as he would his human soldiers. If he is unable, there might come a time when responsibility becomes so dispersed that no one or nothing is accountable. Our notion of the laws of war may fall like a bridge without its tension wires.

The only thing able to resuspend our sense of the laws of war then becomes the advent of an intelligence able to reclaim accountability, either with a reintroduction of humans into the fray or through the dawning of an AI able to pass the Turing Test. Either way we will have progressed or regressed to where we are today. We will once again deal with fallibility on the battlefield in a seemingly deterministic world where our soldiers—human or not—exercise free will and intentionality. This line of reason lends itself to the hypothesis that the human condition, regardless of the elementary basis of "life," just might be inevitable. The end state of an introspective intelligence, no matter how hard we try to outgrow it, may be an infinite regression in dealing with our moral decisions and whatever they may bring.

Though science advances by experiment, its meaning and effects must still undergo debate. Yet we find here a quandary. Our deterministic or "randomly probabilistic" world, obeying the physical laws revealed by science, once again clashes with our own free will. We *choose* to experiment, and ironically the only way to test the "human condition in infinite regression" hypothesis is to do as Hall suggests—build the AI and see what we get. Let's hope we will still have the sense to debate its use and the ability to control its advance. Since I invoked J. Robert Oppenheimer in the beginning of this chapter, I will leave him with the last word. It is a lofty message on freedom and science, but there is warning in the double entendre: "There must be no barriers to freedom of inquiry. There is no place for dogma in science. . . . And we know that as long as men are free to ask what they must, free to say what they think, free to think what they will, freedom can never be lost, and science can never regress."[75]

9

Radical Responsibilities

We are not the Duke of Sung, and we have no use for his asinine ethics.

—Mao Zedong[1]

In the sixth century BC during the feudal wars that defined ancient China for centuries, an obscure lord refused to prosecute a tactical advantage over a much larger force, eventually dooming his army to retreat and loss. As he stood across the river from his opponents and watched them haphazardly wade through high water, his ministers pleaded with him to attack the vulnerable foe. He refused. When his enemy reached his side of the bank but were still too dispersed for proper battle, his ministers once again urged him to action. He again refused, waiting for the opposing force to ready itself for battle before commencing the inevitable attack. In the ensuing battle, the lord was personally injured, and his army ran in disarray. When blamed for the loss and asked why he had not taken advantage of his tactical situation, he responded, "The superior man does not inflict a second wound, and does not take prisoner anyone of grey hairs. When the ancients had their armies in the field, they would not attack an enemy when he was in a defile; and . . . I will not sound my drums to attack an unformed host."[2] This is what led Chairman Mao to resurrect the memory of the Duke of Sung centuries later and dismiss his sense of honor.[3]

We may also find the duke's actions inexplicable and unforgivable. As noted, the duty of a military is to win wars. We would probably judge harshly a general who today, when offered an opportunity to destroy an army's fighting capability, refused to do so. We might snicker at his antiquated sense of chivalry. How could we allow a man or woman so clearly out of touch with the reality of modern war to make such a grave blunder? But we should be careful not to judge so quickly, for though we are not the

Duke of Sung either, his "asinine ethics," to our enduring credit, permeate our operations. In the end they counter an argument, based on misplaced moral urgency, for weapons of impunity in the name of military necessity and the blind pursuit of victory.

The Blind Pursuit of Victory

There is a tension between winning and fighting well that is not understood nearly as well as it should be among all those responsible for the defense of a nation. Some may argue this is just an interesting academic question having little bearing on the real world. They may say we follow the laws of war and the ROE built from them, after all. We play to win, but we do so within the bounds of the law; therefore, whether there is or is not a tension between fighting well—whatever that may mean—it is irrelevant. It is covered by the law and our own regulation. Well, we know about the limits of law, but there are several problems with this line of reasoning that may have implications for the drive toward robotics in warfare.

Unfortunately we do not always operate within the bounds of the law. We are fallible, and our fallibility has consequences. It is also likely to be an enduring part of war regardless of whether we have silicon-minded stand-ins or not. More importantly, our misunderstanding or failure to acknowledge this tension between fighting well—for which, by the way, we have a very clear idea of its meaning—and winning affects the tenor of our debate on all forms of issues surrounding the resort to war, the weapons we use, the way we conduct ourselves, and the manner in which we are able to return to relative peace. My Special Forces colleague from the previous chapter summed up the sentiment, a result of our common ignorance of the profound context of this tension, during a recent conversation. Following my questioning the status and validity of a UAS strike from seven thousand miles away, he said (paraphrased), "You know what I'm going to say: I don't care where it comes from. The effect is the same, and I'd rather it be that than send in one of my soldiers."[4] The mantra is "whatever it takes" to get the job done. It reminds us of the Army tank commander's response to rifle fire from a town. But even here we are fooling ourselves. There are clearly limits to our means, but this tension, and our dismissive attitude toward it, may subvert those limits.

The tension between fighting well and winning is the tension between our lofty ideas about the justice of our cause and conduct, and their slow dissolution at the hands of a creeping moral urgency for victory, as Michael Walzer has pointed out. What good is the former if you are unable to secure the latter? A war in which virtue is more important than victory seems "a very unimportant war." The duke's minister's argument carries the tension to its logical extreme. He states, "If we grudge a second wound, it would be better not to wound at all. If we would spare the grey-haired, we had better submit to the enemy." If we cannot play it our way to win, we take our ball and go home. It is the ghost of Vietnam once more, and the Special Forces officer's argument above. "Fight all out or not at all."[5]

Once engaged in war there is a pressure to do what is required to win, and it gets more intense the longer the war drags on, the more soldiers who are lost. There arises a move against the war convention and toward "particular violations of its rules."[6] The argument of military necessity rears its head again but in an arena where it should not belong. Its requirements, recall, reside in *jus in bello* and are already accounted for in the proportionality requirement. Here, though, it is attempting to justify actions because of some moral urgency for the outcome. If rules are broken in this context, they are broken for the cause. Walzer says that "it is with some version of the argument for justice that the violations are defended."[7] We attempt to defend unjust means by their perceived necessity to achieve the just cause.

The logical conclusion of the moral urgency for victory is the argument that ends justify the means. As an example, the so-called enhanced interrogation measures used on detainees in what was then called the Global War on Terror have been described as torture. In former president George W. Bush's first interview since leaving office, he said he had personally approved those methods, adding, "Using those techniques saved lives. My job was to protect America. And I did."[8] Whether we agree on the characterization of waterboarding or other methods as torture, and whether we can say these were violations of basic human rights, matters less for our purposes here than the clear fact that the president saw them as outcome-based actions. For him, the ends justified their use. This is *usually* an argument senior civilian authorities and military commanders would be unlikely to make explicitly, yet arguments that lead to it—fight all out or not at all—are made with regularity.[9] This is evidence enough that we tend to be ignorant of the nuance of the tension between just conduct and desired outcomes, but again the important piece is how it affects the discourse on conduct and weaponry.

Doing all we can in order to win is often the justification for fielding new weaponry, yet when we make that kind of argument we fail to acknowledge the experience of history. We do not, as a matter of policy, fight all out to win. Arguments based on this assumption and the procurement decisions they lead to are fundamentally flawed. If there was no tension between fighting well and winning, we would still maintain a biological warfare capability. We would support the first use of chemical and nuclear weapons. We would, were it not for the combination of this tension and the directional nature of the law, support bombing civilian populations. We would likely have considered the use of some or all of these means in the border regions of Afghanistan and Pakistan, perhaps many years ago before the supposed world-altering events of September 11, 2001. We do none of these. Yes, they are extreme examples. Some might charge they are absurd. Sometimes projecting toward absurdity is the only way to accurately describe our way of war—we do not fight all out only to win—and counter those who dreamily wish it were not so.

There are other examples not quite so extreme. In the fight for Afghanistan, an unmanned aircraft's sensors were tracking a group of approximately fifty known Taliban fighters who seemed to be standing in the open doing nothing. They were valid targets. To the best knowledge of the crew and the strike authority, they were enemy combatants who could be targeted regardless of their current activity. Recall that combatants are valid targets even when they are not actively engaged in hostilities. Firing on them would have reduced the numbers of Taliban fighters and would have been an incremental victory in that they would have been neutralized from battle, possibly saving the life of an American soldier during a future altercation. Yet in this instance they were akin to the minister's "grey hairs." The strike authority believed them to be attending to the burial of a comrade in what appeared to be a cemetery. ROE may have precluded firing, perhaps under the prohibition against attacking cultural sites, or the strike authority may have taken it upon himself; either way, no clearance to fire was given.[10] Some fifty Taliban lived to fight another day, and somewhere the Duke of Sung was smiling.

We cannot, in good conscience or even with scholarly consistency, make the argument that we have to fight, procure arms, or treat prisoners and noncombatants in any way necessary in order to win. The advent of a technology that aids winning does not create a moral imperative to acquire or use it as would seem to be suggested in "all out or not at all" arguments. These arguments ignore history, the directional nature of the laws of war, and the thing we sometimes lose sight of under the spell of the siren song of technology—*how* we win actually matters more than winning alone. It is a seemingly radical idea when printed in black and white, but it is true. The Duke of Sung's asinine ethics are, as Christopher Coker says, "essential to what Clausewitz called the nature of war [and] the demands of the hour do not require a radical rewriting of the *jus in bello*."[11] To the contrary, the demands of the hour—persistent conflicts in wars among the people—only serve to strengthen them.

The tension between fighting well and winning bequeaths to us a number of "radical responsibilities"[12]: for our soldiers' well-being weighed against their wise use when we must choose to expend them, for the way we relate to our enemies in combat, for the civilians our actions affect, and for the power of our example, based on our first principles, even in the fury and brutality of war. These radical responsibilities, in proper balance, allow us to fight a meaningful war. They allow us to seek meaning in the striving; they recognize our ethics are forged in the cauldron of human fallibility. They point, in the end, to the practical limit of killing—even in just wars—and serve as a warning about our ultimate effectiveness to do our nation's bidding.

Radical Responsibilities

In *Arguing about War*, Michael Walzer describes what he calls the radical responsibility stemming from just war theory. It is the idea that political and military leaders are responsible for the well-being of their own people but also for the well-being of innocents on the other side.[13] It is an officer's lot to weigh these two responsibilities in light of the inevitable damage that comes with the inauguration of war. The responsibility for one's own soldiers is hierarchical. It extends both up the chain to superiors and citizens for execution of the mission but also downward to soldiers for their well-being and for not needlessly wasting their lives.[14] The responsibility to innocents is different; it occurs in a single direction and is one-dimensional. It is outward and nonhierarchical.[15]

The soldiers understand their well-being is conditional. Their officers care for them—in fact are morally bound to them—in one dimension, but their ultimate purpose is to be thrown into battle where normal ideas about well-being no longer make any sense.[16] In this dimension the commander is responsible only for not "wasting" them, for surely some will be killed. The outward responsibility for the "lives his activities affect" is unconditional.[17] Innocents are immune, and to the extent they may justifiably be killed under the doctrine of double effect, the officer's outward responsibility must be equal to or greater than the hierarchical one.[18] Our outward responsibilities cannot be subjugated into our hierarchical responsibilities without reducing the killing of civilians to a calculus of being required in military purpose. Civilians then would be subjugated to the level of our soldiers and, in justifying their killing while protecting our own lives, further subjugated below our soldiers. They would be subsumed into a hierarchy where they do not belong, for they are the citizens of other states "who have no place in this hierarchy."[19] This would have serious implications for their immunity.

Their immunity would dissolve along with the second-tier proportionality required under the doctrine of double effect.[20] At the same time, military necessity would break free from its constraint under proportionality to stand on its own. The moral urgency for victory would then finally wipe away any consideration of noncombatant status as their killing could be justified solely for military purpose. We have traveled that road before in incendiary bombings and nuclear attacks and found it lacking. The separation of hierarchical and outward responsibilities keeps us from traveling the same path—at least for now. But in our robotic age, in the age of persistent conflict, Walzer does not go far enough. We have other radical responsibilities as well—to enemy combatants, to our own noncombatants facing a risk inversion, and to the world at large by the power of our example. I have discussed how we must relate to our enemies, so I turn once again to the risk inversion and our own combatants.

Walzer's outward responsibility for the other side's noncombatants is similarly valid for our own. The fact that, even as recently as 2004, Walzer does not include them directly in his ruminations except as part of the upward hierarchy and the officer's duty to the citizenry, is telling. As I have argued, we have not been confronted with considering their well-being in Western warfare since the end of World War II. We have grown accustomed to this framework to the point we do not even think of them in great works on the philosophy of war. Our noncombatants have grown accustomed to their seemingly impenetrable sense of security, and they are farther separated from the military that fights in their stead. The risk inversion, the transference of risk from combatants to *all* noncombatants at the hands of unmanned warfare, is likely to change all that.

If we cannot subjugate the lives of innocents of other nations into a hierarchy of responsibility involving our soldiers' lives—and the argument is logically and morally compelling—then how could we justify doing the same for our own noncombatants? In fact, there is no moral basis for separating *any* noncombatants, and with this recognition I have perhaps been too hard on Walzer's argument. He might rightfully make the case that his radical responsibility does indeed cover them, but there is still room to explore. Though all noncombatants are morally bound, I have separated them here based only on their fortune or misfortune to live in a society of a given technological state and in a given geographic region either near or far away from the effects of war. This separation is what allows us to see the ramifications of our decisions about employing with impunity and the postulated risk inversion it creates. It seems clear, whether we view our own noncombatants as separate from those we tend to fight among or not, that we can do nothing but accept their immunity on equal grounds. We must view our radical responsibility to them as nonhierarchical in nature, as an outward vector occupying only one dimension. The twofold result of this view is striking.

Our outward responsibility to our own noncombatants first means any actions we take that raise the risk to them, even as we enjoy less risk in the military charged to protect them, ought to be unjustifiable. Second, it raises the prospect we might one day have to justify killing our own citizens under the doctrine of double effect. One can argue we have already crossed that threshold under the auspices of homeland defense in the arena of hijackings.[21] I will not discuss the latter (see note), as the former is more appropriate to our discussion.

It seems unjustifiable to engage in activities that raise the overall level of risk to our noncombatants while the military establishment seeks impunity. It is exactly the same argument made previously on the morality of risk and it questions the moral standing of our search for full impunity. Walzer says there is a moral tension between these hierarchical and outward responsibilities, between those to your soldiers and those to the civilians your actions affect. It means we have to "insist upon the risks that soldiers must accept and that their officers must require. [This will] make soldiering even harder than it is, and it is already a hard calling. But given the suffering it often produces, it cannot be the purpose of moral philosophy to make it easier."[22] To the extent robotic warfare may transfer the risk to all noncombatants, it is morally tenuous to argue for its procurement and use in lethal combat operations. Of equal concern is the quality of our example and the precedent we set.

No matter how much we may wish it were not true, we cannot deny much of the world looks to the United States and studies its ways. Americans are like so many professional athletes, exclaiming loudly or sobbing through tears that they never chose to be role models. The world cruelly thrust this status upon them right about the time they signed their multimillion-dollar contracts. We in the defense establishment may have qualms about our exemplar status, but we have no choice. The 2010 *National Security Strategy* explicitly states, "This strategy recognizes the fundamental connection between our national security, our national competitiveness, resilience, and moral example."[23] This is on page 1.

The United States spends more on defense than the next four to seven countries on the list combined.[24] It is a leading arms exporter. With those arms come good old American know-how and an ability to operate more seamlessly with U.S. forces. All of the armed services' major commands, combatant commands, and unified commands have security cooperation programs whereby members of the U.S. military teach and mentor personnel in other nations' services. We lead by example in these positive instances, but what of our move toward robotic systems of war? There is the simple answer, and then there is a deeper consideration. The latter teams with the lagging nature of the law and may be problematic for the DOD's desires for these systems.

The simple answer, and one that plays well in the media, is that we affect our standing in the world and create more terrorists in the process by use of our armed unmanned systems. The sentiment is summed up by opinion writer Nat Hentoff in *The Trentonian*: "The increasing reliance on pilotless drones operated from afar is a slippery slope that can do serious injury to our standing in the world, not only among our allies but also by becoming a boon to terrorism recruiters."[25] The first part was previously raised in discussions about the conversation between enemies on the battlefield. The second directly challenges the criterion of probability of success with respect to the use of these kinds of weapons. It is important, but I have argued the points in detail already. What we have not discussed is the effect it may have on us and *why* it impacts the power of our example.

Earlier I highlighted the pitfalls of the blind pursuit of victory. In using armed unmanned weapons, we are making a poor utilitarian argument by placing the well-being of our soldiers over the well-being of combatants and noncombatants alike. We are doing so by saying we are being militarily effective—that the good we are doing outweighs civilian deaths and even previously conceived moral notions about how we should fight. But Walzer warns against these kinds of utilitarian arguments in justification of putting noncombatants at risk. He discusses the firebombings of Germany and the use of nuclear weapons in Japan during World War II and says we must move away from the utilitarian argument; it is not sufficient: "There is much else that we might plausibly want to preserve: the quality of our lives, for example, our civilization and morality, our collective abhorrence of murder [the term here denotes the unjustified killing of civilians], even when it seems, as it always does, to serve some purpose."[26] One may consider it preposterous to think a Hellfire missile launched from Nevada and impacting in Pakistan could encompass "our civilization and morality" within the "frag" pattern, but it is not inconceivable.[27]

The moral reality of war is not the same for us as it was for Genghis Khan.[28] We would say it has progressed, but there is no guarantee of its vector toward more restraint, restriction, and recognition of moral equality. Our norms change. If they didn't, neither would our laws, as they are inextricably linked. We have not always changed for the better. There have been major aberrations along the way. Mass bombing of civilians in the last "good" war is but one example. As Christopher Coker says, "If our intention is to influence other cultures . . . then we must remain true to our own first principles—not because those principles are universally true but because they are true for us."[29] Our intention clearly is to influence other cultures. If it were not, the power of our example would not likely be listed on the very first page of the first Barack Obama administration's *National Security Strategy*. We ought to tread very lightly in considering a way of war that may upset our moral norms. If our first principles are no longer true for us, they will be even less likely true for the rest of the world.

There is a very contentious issue in the theory of just war over what is known as exceptionalism or extreme emergency. The ethics of exceptionalism means certain things remain outside the law but are defensible during national emergencies. Gross says that "exceptions violate rules. They are *prima facie* unlawful acts that require perpetrators to defend themselves after the fact," and the best that can happen is society can excuse the actions.[30] The Nuremburg Trial of the German submarine commanders is an analogous situation. They made exceptions to the rules due to the undue burden of risk, defended their actions, were found in violation, but were excused because they had acted as their counterparts had and punishing them would serve no greater good. Exceptions elicit responses of excuse and defense, while rules elicit justification.[31] I do not wish to debate the concept of exceptionalism, because I could not do it justice, but it highlights another issue with respect to exemplar status. Sometimes the exception becomes the rule.

There was a time when the United States condemned—maybe that is too strong a word—there was a time when the United States *counseled* Israel on its use of targeted killings of terrorist leaders in the occupied territories. The Israeli advocate general's explanation for targeted killings once included the terms "exceptional and extraordinary cases."[32] This is the "language of exceptionalism." But following the events of September 11, 2001, the language (now in the United States) was much more a language of justification. Explaining targeted killings in this manner may "reflect emerging norms of warfare."[33] The use of targeted killings in this example is not an argument for or against—that is beyond the scope of this work. It only serves to highlight the possibility of the shifting acceptance of certain means. If the United States can take up the example of the small nation of Israel, what will be the result of it taking up armed unmanned warfare? With the full backing of the largest defense department and the most respected justice department, fielding armed unmanned machines may be at first excused and then justified. And it will validate the use of such weapons for every other nation on the planet.

Moral agency and our deep-seated notion of fighting well create in us radical responsibilities that may challenge our desires for robotic weapons. Our responsibility to our soldiers clashes with the inevitable need to expend them in battle, and this clash leads us to look for other means to pursue warfare. But we are not beholden only to our soldiers. We have other, outward responsibilities as well. We have a responsibility to fight with respect for our enemies, for it, along with our responsibility for noncombatants on the other side, lays the foundation for peace. Noncombatants are morally bound and cannot be separated in terms of how we deal with them in war. Therefore, we have responsibilities to our own noncombatants to shoulder the risk in war. Methods and means that invert the risk and transfer it to noncombatants of either side cannot be justified.

Finally we have a responsibility, through fighting well, to extend our example. We must guard against validating ways and means of war by our own actions. In doing so, we attempt to maintain our first principles even though it makes our soldiers' chosen calling more difficult than it already is. In doing so we seek ways to address our fallibility and maintain meaning in war.

Fallibility, Ethics in the Striving, and Meaning in War

In chapter 3 we discussed decisions made in combat in the passion of kindness, decisions involving not just doing what is morally required, but less than is permitted.[34] Unfortunately other passions often interfere in combat operations, and their results depict the darkest side of war. Yet even here there is room for hope. There is also room for caution as we proceed farther into the robotic age of warfare. In the first case, there is the realization that our ethics are produced only in the striving for better ways to live and relate to others. Our personal quest makes us better persons. Our nation's quest, most visible in our willingness to submit to the rule of law, adds to the power of our example. In the second case, robotic warfare may deny us the human sense of striving for a better ethic in war, denying us meaning and affecting the warrior ethos. It may affect our conversation with our enemies and those we would like to influence due to the ambiguity about responsibility and accountability for lethal autonomous operations, especially when things go wrong as, unfortunately, we know they will.

The technological trend in war allowing us to attack from greater range primarily due to increasing precision paints a picture of the possibility to conduct war without error. It is a step beyond what Martin Cook called "immaculate war" in an earlier chapter.[35] Immaculate war is war without casualties, and we have addressed the fallacy and pitfalls already. Error-free war is without mistake. It removes the fog and friction, allowing perfect action at all times. As Michael Ignatieff says, "Legal war, when linked to precision weaponry and targeting, creates an expectation, which military, public and politicians alike come to share, that war can be clean and mistake free."[36] It is a myth. It was destroyed, or should have been, with the mistaken bombing of the Chinese embassy in Belgrade during Operation Allied Force.[37] Even with precision weaponry, the most accurate satellite photos available, the best maps, and the efficient operational planning process developed in the U.S. military over many decades, human error still intervenes to shatter the idea that Clausewitz might have been wrong. Analysts apparently classified the embassy as something else. It was a tragic mistake. Civilians lost their lives, but it was not a crime of passion in war. The intelligence analysts did not do what they did purposefully or with any malice. Though the end is the same—the deaths of innocents—we somehow view it differently when the killing is more personal.

One of the studies Ronald Arkin cites in his discussion on the possibility of robots to act more ethically than humans is a report done by the Mental Health Advisory Team for Operation Iraqi Freedom, which studied the mental well-being of deployed soldiers and Marines. The advisory team also studied what it termed "battlefield ethics," and this is where Arkin found fertile ground for improvement. The team surveyed soldiers and Marines about the treatment of noncombatants in Iraq and how they should be viewed. Four percent of soldiers and seven percent of Marines reported hitting or kicking noncombatants (this may include civilians or previous combatants now hors de combat) when it was unnecessary.[38] Eight percent of soldiers and nine percent of Marines reported modifying ROE in order to accomplish the mission.[39] Five percent of soldiers and seven percent of Marines reported ignoring ROE in order to accomplish the mission.[40] Only forty-seven percent of soldiers and thirty-eight percent of Marines thought all noncombatants should be treated with dignity and respect, and seventeen percent of both thought all noncombatants should be treated as insurgents.[41] Perhaps most disturbing, fully thirty-seven percent of soldiers and thirty-nine percent of Marines thought torture should be allowed to gather information about insurgents, and the numbers rose to forty-one and forty-four percent, respectively, if the information might help save the life of a comrade.[42] Clearly there is room for improvement in our passionate decisions. Yet even when our soldiers step into the realm of crimes in war, there is room for the progression of ethics and knowledge.

Crimes committed in the heat of battle, or its immediate aftermath, have probably been a staple of warfare since it first began. It is hard to engage in a dark art involving the killing of human beings and have it play out without both the passions of kindness and rage. We have seen the cold and calculating murder of millions under state genocidal policy, such as the world witnessed in the Holocaust, and the ethnic cleansing and mass murder in such places as Srebrenica and Rwanda. Killing from the passion of rage in time of war represents the descent of warfare into its most barbaric catacombs, but because of the laws of war and the moral foundation they rest upon, we are not destined to remain there. The *New York Times* reports, "Over the last nine years, as the Army has cycled hundreds of thousands of soldiers through combat duty in Afghanistan and Iraq, it has also court-martialed 34 on murder or manslaughter charges in the killings of civilians in those conflict zones. Twenty-two were convicted, and 12 acquitted."[43]

A senior commander in Afghanistan in recent comments to a National Defense University audience said that when we make mistakes and kill innocents, we immediately bring it to the fore, apologize publicly, learn how to attempt to prevent it in the future, and move forward.[44] When we find alleged crimes, as the *Times* quote clearly shows, we prosecute, eventually holding those responsible accountable for their actions. Our responses to the darkness of humanity in war are powerful indicators of our recognition that our sense of ethics, even in the depths of war, is bettered in our striving. Martin Cook explains:

> The rhetoric of a strong moral basis of the military profession should be taken as a testimony to real and powerful aspirations—and aspirations to be deeply valued in the society. The value of those aspirations remains and should be honored, however far short of them we sometimes fall in experience. These aspirations are the foundation of the military virtues that preserve and sustain some of the noblest of human values: to serve others even at the cost of personal sacrifice, and to discipline one's mind and body so that it serves a purpose larger than self.[45]

We have to ask ourselves, in scholarly debate removed from the emotionally charged atmosphere the context of the question lends itself to, whether we are willing to trade the passion of kindness we find in warfare for the cold and heartless killer that a lethal autonomous system will undoubtedly be. In the former we find the progression of our ethics, including the power of the example the journey projects. In the latter we may create a system that will not kill a noncombatant through the passion of rage. It will also feel no qualm about killing a soldier escaping the desolation of war for just a moment by leaning against a foxhole wall, lighting up a cigarette, and thinking of other, more peaceful things.

If error-free war is a myth even now, will we be able to build a machine that can do better than humans on the battlefield? Arkin thinks we can. I am not so sure. I was once that young lieutenant captured by the challenge of figuring out why an IR missile was not hitting its target. The "consecutive miracles" of modern weaponry often fall short. Robotics must overcome sensor issues, wicked AI programming problems, and fragile software code. If it happens (and I fully admit it may), will the dispersal of responsibility affect accountability to the point where we will not be able to learn from the inevitable mistakes our lethal autonomous systems will eventually make? Will we be unable to say we found the one responsible and convict them of their crime? How will robotic warfare affect our ethical striving? Christopher Coker asks it this way: "In hollowing out the warriors' honor, or threatening to replace warriors themselves with machines—we risk ignoring Aristotle's insight that ethics is in the striving. In the striving we discover much about ourselves and otherness. If we are not allowed to strive, what will there be left to discover?"[46] We may not like to admit it, but human fallibility plays an important role in the experience of war. Without it, war will be less real. Without it, it will lose much of its meaning, and warfare without meaning becomes nothing more than awful waste.

The Practical Limit of War

Perhaps I have not given due consideration to those with no time for "radical responsibilities" and the academic theories of war. The colonel in the Pentagon, thinking war an amoral exercise, likely has no need of my "asinine ethics." There is but one real measure of effectiveness in warfare, after all: the successful conclusion of victory, the attainment of the political objectives that sent us off to battle in the first place. There is no question that I would be remiss in leaving this aspect out of any discussion on a possible new way of war. Considerations of such things, including the personnel and organizational transformations required, ought to increase the combat capability of military forces and allow for the desired political end state. Otherwise, why pursue such means in the first place?

In chapter 5 we discussed some of the issues surrounding impunity and the politics of war whereby a great concern is what "unusually usable" weapons may do to the civil-military relationship and the character of military advice to civilian authorities. There the issue was how such weapons may limit the constraints on warfare, causing a resort to war in instances when such a decision might not have been made in the past and in the absence of weapons of impunity. In chapter 7 I discussed the problem of employing a means of war that undermines the just war criterion of probability of success based on the inability of the enemy to relate to the means. It may destroy the respect essential for building a lasting peace. Here, though, we need to discuss another dimension of the probability-of-success argument, this time in relation to public and political will, which adds an interesting twist on the "usable" weapons concept. If combatants experience a resistance to killing in war, there is perhaps an equally powerful resistance on the part of the public to continue killing when it appears to be in the absence of some semblance of a contest, when it is viewed as wanton or simply too easy.

In the last days of Desert Storm, as Iraqi troops were fleeing Kuwait in large convoys toward Basra and beyond, air forces of the coalition pounded the seemingly defenseless Iraqis. The intent was to damage Saddam's army to the point it could no longer threaten its neighbors. The fight seemed to have gone out of the Iraqi force, but there had been no suing for peace as of yet. The Iraqi equipment and those who manned it were, by any reading of the laws of war, legal targets. The road, which became known as the Highway of Death, was what we military aviators might call "a target-rich environment." But the previously discussed CNN effect was in full force as scenes of destruction were beamed to living rooms around the world. As pilots described the actions along the highway as "shooting fish in a barrel," the public began to wonder if it was all really necessary.[47] Gen. Colin Powell said, "The reports make it sound like wanton killing."[48] As discussions began about a ceasefire, powerful images of a superior force having its way with an army in retreat weighed heavily on the minds of decision makers in Washington. The debate became "uncomfortably intense," with two of the staunchest members of the coalition publicly beginning to question when the war might end.[49] Though Gen H. Norman Schwarzkopf, the combined forces commander, thought many of the Iraqi troops had abandoned their vehicles, fled into the desert, and been spared a fiery death, in the end the only thing that mattered was how the public viewed the destruction along that four-lane strip of asphalt.[50]

By most accounts Desert Storm was a just war, one fought in the face of the crime of aggression on the part of the Iraqis. It was, also by most counts, fought with justice. On the Highway of Death, it appeared that combatants met combatants in legitimate acts of war, yet something was still amiss. I have argued the law is not enough to justify new weapons and ways of war, but here it appears justice is sometimes not enough either. In speaking of the end to the Gulf War, Walzer says, "Justice is not the whole of morality. One may object to killing in war, even in just war, whenever it gets too easy."[51] Michael Ignatieff says: "Put another way, a war ceases to be just when it becomes a turkey shoot."[52] Finally, again from Walzer to close the thought: "A 'turkey shoot' is not a combat between combatants. When the world divides radically into those who bomb and those who are bombed, it becomes morally problematic, even if the bombing in this or that instance is justifiable."[53] We have discussed the morally problematic nature of impunity in isolation before, but here we see its very real effects on a public whose military forces are engaged in the field.

If we reel from images of aircraft leaving a swath of destruction across the landscape in a just war, it seems logical to conclude we might have the same reaction in seeing our highly efficient and unfeeling killing machines ripping through our enemies' formations with calculated abandon. Who, while watching scenes—even fictional ones—of doughboys climbing from trenches only to be cut down in their first steps, has not felt their stomach sink? Our feeling of despair for their fate rises as they ascend; we know they have no chance. It would likely be no different in robotic warfare, and surely every robot "warrior" would be equipped with high-definition cameras to allow the generals to view the action from their command centers, just as they do now in watching Predator video feeds. There would be no shortage of footage for those with either a twisted sense of voyeurism or a genuine sense of awe.

It seems possible that the morally problematic reality of such efficient killing might push war to real pragmatic limits. It is possible, if history is any guide, that such killing might open our eyes and force us to consider the blind pursuit of victory yet again. We might find that fighting well does indeed matter and that heartless machines no longer fit our views of what, exactly, that means. We would then be left with a hollow force of weapons we were no longer interested in employing, even in the pursuit of just ends. Coker says, "Looking at the post-human future that some Pentagon planners envisage, one has an unmistakable feeling that they have little real understanding of the imaginative and psychological price we may all pay for what they would like war to become. The ultimate challenge of the robotic age is that the robots will inhabit a world of means largely divorced from ends."[54] We would have military means that could not, due to our own moral character, meet desired political ends. Ironically, we would have reached the practical limit of our means of war. We would be left with unusually useless weapons.

Anything but Asinine

In order to maintain meaning in war, we have to recognize the seemingly counterintuitive prospect that how we win is more important than simply winning. The moral tension between fighting well and winning arises when we attach a moral urgency to the ends at the expense of the morality of the means. Arguments for procuring and using the latest in robotic technology in the name of "fight all out or not at all" are historically and logically unsupportable. We do not now, and probably should not, fight all out simply for the ends. Such arguments ignore the current reality of war among the people, where setting the conditions for the ends may be the only thing even possible and the ends are far from certain. The only way to prevail in this environment is to understand that fighting well *is* winning. The duke's asinine ethics, as it turns out, are essential.

Our moral agency and our claims of justice in warfare bind us to four radical responsibilities. We have a hierarchical responsibility to our soldiers and superiors. We must look after the well-being of soldiers, and we must not waste them in battle. At the same time we are bound to our superiors and citizenry for the execution of policy by force. Our responsibility to *all* noncombatants is nonhierarchical. It is an outward responsibility equal, at least, to our hierarchical one. It is morally unacceptable to put our own safety above that of the noncombatants whose immunity we must protect. We have a responsibility to converse with our enemy as equals and with mutual respect. This and our responsibility to noncombatants set the foundation for peace and are particularly important, it seems, in wars among the people. Finally, we have a responsibility for the power of our example. We must operate in concert with our first principles, ever mindful of how the ways and means of war will be perceived and what precedent they may set.

Lethal unmanned and robotic weapon systems upset our radical responsibilities by transferring risk to noncombatants, altering our conversation with our enemies (possibly destroying our moral equality), and validating weapons of war that may change the norms of warfare. Should these occur, we will lose the meaningful basis of war and our ability to influence it by our own example in seeking war among equals who consent to battle, fight with justice and for just cause, respect the immunity of innocents, and work toward a lasting peace founded on respect in battle. The law lags our introduction of new weapons of war. This lag may be diverging due to the pace of technological change, and we may end up struggling with the unforeseen, attempting in vain to rein in our robot "warriors." Again, nuclear weapons are an instructive case. We excused their use, then justified their being, and in doing so gave room for every other state to do the same. We have spent the last nearly seventy years trying to get Pandora to close the box. As Oppenheimer said, not meaning it as the warning it is, science does not regress.

As science progresses, it may create an impotence in the ability to achieve our objectives in war. A military capability is not a capability if it cannot be wielded toward some justified end. In that regard, it exists only to bring about a desired political end state that cannot be gained by any other means. Procuring weapons that, by their very use, destabilize the just war criterion of probability of success seems like folly. Whether because our enemies refuse to submit to the desired effects of our use of force based on their own views of fighting well, or a public pulls back from the wanton destruction and supremely efficient killing of modern weapons, the result is the same. Our implements of war, no matter how technologically advanced and protective of our soldiers' lives, may cease to be effective.

I opened this work with Kant's categorical imperative. It is often thought to be encased in the Golden Rule: Do unto others as you would have them do unto you. Yet there is something lost in this summation. The categorical imperative compels more than simply to act so as to be treated equally well. Kant takes it immeasurably farther: "Act as if the maxim of thy action were to become by thy will a universal law of nature."[55] Here there is more than act and reaction. There is the altering of fundamental laws of nature. With our progression toward lethal robots, we may be redefining the very nature and meaning of war. If we are smart, we will carefully consider the possibility that the acquisition and use of these systems will become the universal way of war, but having them used against us should be the least of our worries. Our deeply moral reaction to coldly efficient, too-easy killing is an internal voice of caution prodding us to take great care. If, by our own example, heartless weapons become the norm in war, the only way to make them useful again is to let go our revulsion for the ease with which our machines do our killing. We will have come full circle. Our unusually usable weapons will have become useless in our disgust for their efficiency, but in doing so they will have altered the universal norms of war. Then, in order to restore their usefulness in the violent politics of war, we will have to trade away a portion of our morality in redefining justice in war. It would be an exceedingly high price to pay.

10

Inevitability, Persistence . . . and Heart

Our knowledge of science has clearly outstripped our capacity to control it. . . . We know more about war than we know about peace, more about killing than we know about living.

—Gen. Omar N. Bradley, Armistice Day address, 1948[1]

Let us not become so preoccupied with weapons that we lose sight of the fact that war itself is the real villain.

—President Harry S. Truman[2]

Behold! I show you the Last Man.

—Friedrich Nietzsche[3]

Christopher Coker, in his magnificent book *Ethics and War in the 21st Century*, says the first question a soldier asks is "Why?" The second question, "first posed by Socrates [is] 'How should we live?'"[4] Invoking Socrates, as it often does, leads to salient questions of being. Why should we develop lethal unmanned systems, and further, why lethal autonomous systems? Instantaneous shallow answers, such as safety and risk, jump forward for the former. Answers for the latter are far less direct, perhaps even hedging.

I have argued against both in this work, but I am not naive enough to think it will do much good. Moral issues at the heart of the just war ethic probably will not make the news. These are not the sexy moral issues that tend to splash across headlines. Moral and ethical arguments such as these are unlikely to influence policy, though I can hope they will give some pause. Even the practical arguments, such as military deskilling and the possibility of acquiring weapons the public will render ineffective by their unwillingness to use them, are similarly unlikely to cause much stir. We have been in constant conflict for over twenty years, and as I was writing this, U.S. forces had just entered another theater of armed conflict in Africa. The outlook for the emerging markets of war is bright, and demand continues to grow. It seems we will have to deal with the second question. How should we live in a world of robots?

Socrates for Robots

This work has argued the tenuous moral, ethical, legal, and practical issues of armed unmanned systems, including current operational systems and future lethal autonomous systems. Lethal unmanned systems undermine the foundation of the laws of war by removing the moral equality of combatants. They contest the legal status of those operating these systems. Additionally, they may be ineffective at bringing the current conflict and future conflicts like it—the paradigm of persistent conflict in wars among the people—to a proper conclusion. Therefore, their use undermines the probability-of-success criterion of just war theory and reduces their credibility as just weapons. There will undoubtedly be push-back from industry and certain elements in the services, but these ideas are very unlikely to limit the opportunities for growth of the robotics industry in general and robotics in the military in particular.

There is a maxim, Amara's Law, first stated by futurist Roy Amara, that says the effect of technology tends to be overestimated in the short run but underestimated in the long run.[5] In reading the DOD and service "roadmaps" for unmanned technology, in watching recruiting commercials, in reading the items services tend to release to the press through public affairs offices, one gets the sense unmanned vehicles are seen as *the* sole future of the armed services. Predators and Reapers are winning the war against nasty insurgents in Afghanistan and Pakistan. Robots are neutralizing the threat of IEDs for soldiers on the ground. Robotics is the next great thing. It may be, but it is very unlikely the panacea it is made out to be, particularly for the combat arms. We are very early in our fielding of unmanned systems, and there is much to learn. Peter Singer warns against judging technologies too early for things they do today.[6] He was really chastising the services for complaints about crash rates, system failures, and the service cultural friction that UASs seem to create, but his point has to cut both ways if it is to be meaningful. We also should not judge what they do today as the valid end state. It is entirely possible we have it wrong. We cannot close our eyes to questions about whether their current use, particularly for kinetic operations, meets the needs and overall objectives of U.S. security and military strategy.

Gartner, Incorporated, a leading information technology research and advisory company, has devised a concept to explain the life cycle of new technologies. It is called the Gartner Hype Cycle and is shown on the following page. The height of the peak of inflated expectations is where Amara's Law begins. The plateau of productivity is where it ends.

Unmanned vehicles are riding the wave on the way from technology triggers to the peak of inflated expectations. Nowhere is this more apparent than in the realm of autonomy. The aforementioned research by Missy Cummings is enlightening here yet again. It is showing that we might be concentrating human control and input, particularly for UAVs, in the wrong elements of the control system and attempting to force autonomy where human input is most useful.

We are currently able to kill single individuals through a vast network of communications supporting UAS combat air patrols. Personnel in several locations fly the same UAV (at different times) and fire missiles like they are sniper rifle rounds. Then we claim we are winning. But at what cost? Is what we are doing now with lethal unmanned systems sustainable? How will they fare in a different sort of war in contested airspace? That may be when we hit the trough of disillusionment and we recognize our hype has blinded us to other, likely more mundane future uses of unmanned technology. The only thing we know for sure is the unmanned force of the future will look nothing like it does today and will be doing things no one has yet thought about. These may include kinetic capabilities, but by no means should we expect it simply because it happens to include them today.

There is plenty of room to grow our use of unmanned systems in areas that do not require them to be armed. These may include robotic use in supply, warehousing, logistics, medical support, casualty evacuation, personnel actions, and many other supporting functions we likely have not thought about. Major robotic manufacturers—because of recent technological advances in sensor capabilities—are currently seeing future markets in sorting, handling, and packaging functions.[7] Capabilities like this would be useful in war matériel storage locations and prepositioning ships. Given the moral, ethical, and practical issues involved in lethal robotics, it would be smart to start thinking about how to smooth the curve toward the plateau of productivity rather than beating our chests over our lethal unmanned systems and continuing to inflate their dubious significance.

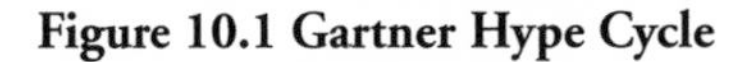

Figure 10.1 Gartner Hype Cycle

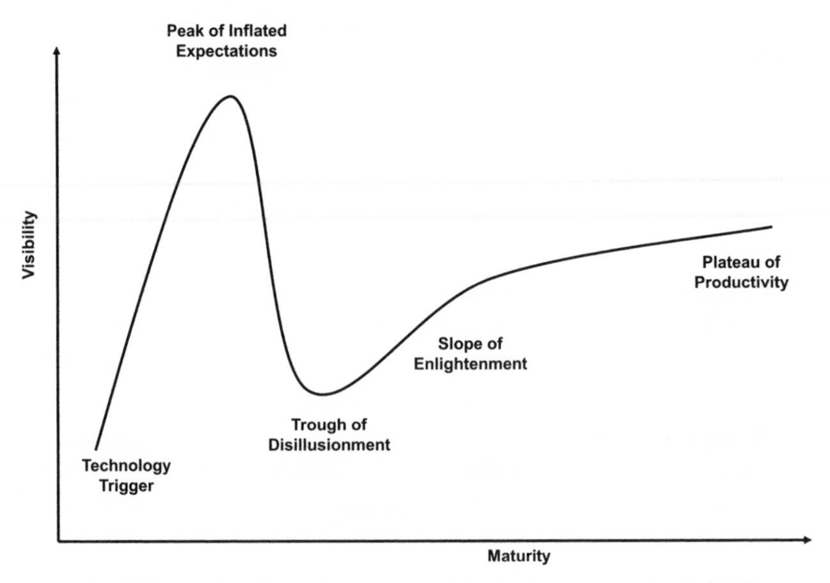

Adapted with permission from Gartner, Inc., http://www.gartner.com/technology/research/methodologies/hype-cycle.jsp.

As we ask "how should we live," we must consider the power of example. The United States should work through international organizations and legal conventions to codify limitations on armed unmanned systems. Unilateral action satisfies our sense of what is right and just, but it will not slow the proliferation of these systems. Because of their ability to upset the current asymmetry on the battlefield, other states and non-state actors will increasingly attempt to acquire unmanned capabilities.[8] This is an opportunity for international law to be proactive, and though it also will not completely stem the tide of armed unmanned systems, at least there will be an international legal regime with which to deal with the issues that may arise. Solving these issues will take a very long time. Some states will find reasons for gaining or maintaining these capabilities and will be slow to move, or abstain altogether, in efforts to restrict the use of armed unmanned systems. There will not be the kind of consensus enjoyed in the lead-up to the Ottawa Accords, but that is all the more reason to start the process now. If we do not, we will once again be left with the lagging nature of the law as current capabilities diverge more and more from constraints on their use.

Citizens of free nations ought to insist on a system of "ethics-in-robotics advisory councils" to advise research agencies, academic institutions, industry partners, and acquisition functions. These might be made up of ethicists, lawyers, engineers, historians, psychologists, philosophers, scientists, and others, with an advisory role on possible ramifications of current or future research. Such councils may have no decision authority, but their inputs would be noted in any grant decisions, acquisition milestones, research proposals, and so forth for those agencies and institutions driving technology forward.[9] As matters of public record, these inputs would likely act as a constraint on the excitement of "pure science." Such records could then be incorporated in future vision documents and "roadmaps" on robotic technologies, continuing to further the important concepts of morality and ethics in the science we choose to do.

There are thousands of uses for military robotics. The defense establishment is pursuing some; others have not yet been considered. I recently asked an executive for a major trade organization advocating for unmanned systems why he was so adamant about fielding armed unmanned systems. He responded with his own question, "Why shouldn't we?" When I proceeded to lay out the arguments in this work, he shrugged them off one by one. When I asked again for his rationale, he had but one: to protect the American war fighter. I asked about all the other markets for unmanned systems in military operations, but they clearly were not enough for him.[10] Granted, advocacy is his job, but his reaction, and those of others in the industry and around the services, is evidence of these technologies' place on the Gartner Hype Cycle. They are riding the wave toward inflated expectations. Unmanned systems, to zealots and businessmen, are all things to all people. They ignore, or are unaware of, Amara's Law. Projections of the effects of unmanned systems are likely being overestimated in the short run, and very few people are even talking about the long run.

The implication for the defense establishment is that it is time to get realistic. It is a very good time to consider the rhetoric and reality of unmanned systems. The DOD has good reason to make policy and define a vision that brings expectations back into check and hedges against disillusionment. Now is not the time to be seduced by the technology. Now is the time to search our own hearts and minds and make moral and objective decisions about the future of robotic systems in military operations.

Robotic technology shows both promise and potential for military use. In a world of persistent conflict, seen playing out in the news on a daily basis, there will probably be a market for military robotic equipment even though the defense budget is likely to shrink in the coming years. If robotics can show cost-saving, manpower-saving, or manpower-reallocation possibilities, there is reason to remain bullish, like it or not, on robotics in war. How we proceed—how we decide to live—will say much about our character and will have far-reaching ramifications for the conduct of war.

The Canvas of History

There is a thread running just beneath the surface throughout this work. It is a contentious issue, and one that is uncomfortable to address, but having come this far we might finally be ready to confront it. It is the idea that if we have or can produce a technological solution to prevent having men and women killed in battle, we are bound to do so. It is the argument that there is a moral imperative to protect the lives of soldiers no matter the consequences and even if it places innocents at greater risk. I believe this is a mistaken and dangerous notion, but arguing against it pits one against walls of black granite, half-staff flags, and weeping mothers whose sons and daughters went away to war and never returned. It is hard to say, in light of these stark reminders, that this is, unfortunately, the terrible cost of war—and it ought to remain so.

I had the great fortune to walk along the now pristine shore of Normandy a few years ago with my family. My mother had an uncle who landed on one of those beaches on June 7, 1944—D-Day +1. He was shot twice but was fortunate enough to survive to tell tales in years to come to anyone who would listen. As our two very young boys ran around the beach drawing lines and shapes with sticks, the breeze caught my wife's hair and brushed it aside. I saw her eyes were welled with not-yet-fallen tears. She looked at me and said, "I'm just thinking of all those mothers' sons." What hit me at that moment was the near silence and complete serenity of that place nearly seventy years after it was stained with the blood of valiant men on both sides. It remains one of the most powerful moments I have ever known.

It is hard to imagine rocking around in a landing craft approaching that shore. The sound of rounds impacting the ramp was possibly the loudest thing those men heard despite the din of diesel engines, waves, and gear banging against the metal deck. Many were likely racked with a fear that could make muscles and minds seize. And yet, when the ramps slammed down, most of them waded toward the sound of the guns. Many never made it through the surf to the beach. It is an awful thing to do to ask individual young men and women to endure such horror. Yet when they are banded together for something larger than themselves, their willingness to do so approaches the sublime.

On top of the cliffs that once housed the fortifications and manifestations of war rests the Normandy American Cemetery. It holds the remains of 9,387 Americans killed in action and lists another 1,557 missing.[11] As separate entities with hopes and dreams of their own, those men were not fighting for freedom or any other lofty goals that day. They were fighting for each other. Their individual acts of courage are worthy of our praise and remembrance, but doing only that lessens their sacrifice by glorifying the ultimate and dirty business of killing in war. It misses the larger point.

The horrible individual act of ordering young men and women to risk all is part of a far larger scheme that communicates what really matters to a nation. The willingness to drive them into battle, to fling them against the canvas of history, says that for which we fight is important enough that we are willing to risk the future lives of our budding youth in order to achieve it. Risking a single life seems tragic. That we are willing to ask droves to face the possibility of death on the fields of strife for some greater good is the only thing that makes war meaningful as a human activity. Without the willingness to risk greatly, we have no business making decisions about others' mortality. Camus was right. We have to be willing to die in order to kill.

Killing with Heart

Armed unmanned and robotic warfare is unlike any revolution in military affairs of the past. There will be those who will argue—in fact, some already are—that "sending in the 'droids" is the same thing as the introduction of the longbow or gunpowder or the aircraft. They will say it is just the next step along a six-thousand-year continuum. They are wrong. The fundamental difference is that we are attempting to remove *all* risk to one side's combatants while transferring it to *all* noncombatants, regardless of geocoincidence or technological standing. That has never happened before in all of recorded history. The risk inversion is the true revolution in warfare, and it is being driven by robotic technology.

The trend toward removing all risk for those engaged in war has serious moral implications. Perhaps most profound for those doing the killing is the removal of the moral basis for their status as combatants, for without the consent to be killed, the moral equality of combatants is destroyed. When this happens, justice in war degenerates. It breaks down the contract between combatants and removes any sense of innocence for their legitimate acts of war. It may affect how they will be treated both during war and after as the rights of combatants, granted on the basis of the above consent, become null and void. Also troubling is the probability of the unjustified killing of noncombatants, invalidating the concept of double effect, due to the unwillingness of combatants to shoulder risk in means that might better protect noncombatant immunity. These issues only come into play once the war has begun. More troubling for the public at large is the effect robotic technology may have on the resort to war.

Armed unmanned and robotic warfare capabilities subvert the last-resort criterion of just war theory in ways that are far deeper than the calculus of risk and reward. They impact the politics of war in interesting ways. Citizens, believing they know more about war than at any time in the past because they have "seen" it on televisions and computer monitors, demand riskless and immaculate war. They do not understand its impossibility, nor do they realize their request only makes the resort to war easier for their elected leaders. They abdicate their responsibility to their republic in the false notion that their own combatants' deaths represent the vileness of war. They do not remember Truman's warning and have somehow forgotten the true villain is war itself.[12]

Armed unmanned systems may lead legitimate authorities to view acts of war as costless because they neglect the universal costs required in proportionality debates. They may even see the use of armed robotic systems as actions other than war, which has legal implications that will likely ripple through the laws of war in decades to come. No one can possibly know how our current actions will later be judged.

Most disturbing, however, is how the quest for impunity may be affecting civil-military relations and the once distinct boundary between military advice on the means and mechanics of war and discussions on whether the resort to war is appropriate. The former is the domain of military leaders, while the latter is that of the civilian authorities. Unmanned and robotic systems may be eroding this division of responsibility as the presentation of "unusually useable" weapons bridge the once separate *ad bellum* and *in bello* decisions. The obvious issue is how these weapons may lower the barrier to war. The more nuanced one is what this erosion may do to our future ability to judge combatants apart from their cause, heretofore the foundation of the combatant's right to be found innocent of his legitimate acts of war regardless of the justice of his cause.

The irony of our desire for impunity in war is that on some level we recognize even this kind of war does have a cost. We talk on the one hand about how seeing death on a video screen somehow sanitizes the situation, yet we are concerned with our "remote warriors'" psychological health after watching such scenes day after day "at war," then returning to their families at night. We know impunity affects those who are doing the killing, but to what extent and in what manner remains a mystery. Impunity affects our "conversation" with the enemy, yet we choose to ignore it or purposefully cast it off as irrelevant, highlighting only the body count of bad guys as the justifiable ends. We are blind to the possibility it is affecting the chances of a just peace. Confronting the effect these systems are having on the reasonable-chance-of-success criterion of just war is lost in the background noise of the moral urgency for victory. Yet without a reasonable chance of success there can be no victory. It is the height of irony, and only our self-delusion allows us to make the veiled "ends justify means" argument when the means themselves may be denying the ends we seek.

In the end, the enemy's ability to resist both the means and our desired outcome may not be as important as how our machines' killing may affect our own public. Armed unmanned aircraft kill with relative ease, notwithstanding the possible effects on their operators. They are currently deployed without public outcry to kill those inside U.S.-allied states. Lethal autonomous robots promise even greater ease and efficiency. All this leads to the possibility of a public awakening to what may appear as wanton slaughter. If the public reacts as it has in the past to similar scenes, it is possible our useable weapons will be driven to impotence. They might become useless in our revulsion to the blind pursuit of victory at the cost of fighting well. We would be left with military means once again unable to achieve any kind of political ends. We might avoid all of this, and the depravity of Nietzsche's last man, by remaining true to first principles and once again cultivating a warrior class and ethos.

William Tecumseh Sherman was partially right when he said, "War is cruelty and you cannot refine it," but even in the hell of war we have found humanity.[13] The fallibility in our humanity sometimes makes us gasp in shock at our capacity for atrocity. Yet at times it makes us understand our ultimate goodness, for even in war we are capable of the passion of kindness. Martin van Creveld says, "War is life written large . . . war alone both permits and demands the commitment of all man's faculties, the highest and the lowest."[14] In striving for the highest, the warrior finds personal meaning in war. This meaning, far more than the law or any other restraint, is what contains war within its moral framework. Michael Ignatieff describes it this way: "But law's dominion over war has always been uncertain. The decisive restraint on inhuman practice on the battlefield lies within the warrior himself, in his conception of what is honorable and dishonorable for a man to do with weapons."[15] Yet even honorable things done with weapons are often hard to do, and here is an important question we must all answer for ourselves. Are we willing to trade away a resistance to killing, which sometimes invokes the Duke of Sung's asinine ethics, for a machine programmed to kill in complete and strict compliance with ROE? Do we want near-errorless lethality, or would we rather have "reluctant professionalism?" We must remember that the law, and the body of ROE stemming from it, is not the whole of morality and ethics. It is important to preserve a warrior ethos if we are to continue, as it looks as if we are sure to do, in the business of war.

Our autonomous robots will be able to kill without emotion. Some tout this as a strength, yet if we were talking about a human we would call him a psychopath. There ought to be a warning there. Charles Dunlap cautions, "Statesmen and soldiers must be especially wary of any actions that might erode the altruistic warrior ethos that underpins instinctively proper behavior in the crucible of war."[16] Christopher Coker says that "we would find ourselves living in a world without meaning, in which taking the life of another person would be no more wrong than unplugging a computer. . . . A robot can inflict suffering, and even affront human dignity, but it cannot be altruistic. . . . It lives, in that sense, in what we would see as a meaningless world.[17] The possibility of killing out of the passion of rage, even in the nonaltruistic darkness of atrocity in war, cannot serve as the sole moral argument for removing the warrior from war or taking the human off the battlefield. Ignatieff also says, "Rules honored more in the breach than in the observance are still worth having."[18] It is so because of Clausewitz's insight, again from Coker, that "ethics is what prevents war from becoming absolute. . . . It is what makes war possible for humanity."[19] But the possibility of war need not—indeed should not—mean its inevitability.

There was a time when American use of force was done out of deep reluctance. It was, for a time, the national ethic on the resort to war, but reluctance seems to have slipped over the past decades. Our nation as a whole is now immune to the effects of war. We pay for it neither with blood nor treasure. It has made us more warlike without having to be more warriorlike. We seem to have lost a sense for the gravity of acts of war; our sense of the tragic is waning. We have forgotten that the resort to war, except in response to existential aggression, represents a failure of the universe of other policy options. When we go to war we fail ourselves, and we fail those who will surely pay the highest price. While sometimes necessary, wars are sure to bring death and strife. We must not shy away from this basic fact. It should be on the forefront of our minds when we commit to war, and in so doing we ought to be willing to risk our own lives. We ought to try to conceive of the staggering human cost for those we will fight and for those we will send to fight. The moral imperative to protect our soldiers at all costs is false. Protecting them from reasonable risk is warranted. It is one of our radical responsibilities, but avoiding their deaths at the cost of others is unjustifiable. Avoiding war altogether ought to be in the greater interest, and means that make it more likely in the end cannot be used as moral justification for protecting those involved in it once these means hasten its eventuality. It is at once circular and mutually exclusive. War ought to come with consequences—otherwise we are truly doomed to repetitive and persistent conflict.

Like the young Italian lieutenant who could not bring himself to kill his German peer as the man blissfully smoked a cigarette, in matters of war we ought to let our humanity intervene. War is a human activity that must remain so. We will lose fathers, daughters, sons, and mothers—this much is true. That is, painfully, the cost and enduring legacy of war. It should serve to make us hesitate for just a moment, if only to try one last thing, before committing to a price we can never fully afford. Warriors understand and have freely chosen their lot, for the warrior path is one walked only in deliberation. They will do our bidding when we ask it of them. We must be willing, as hard as it is to reconcile, for them to risk the finality of death if we are ever to expect them to kill with heart.

NOTES

Preface

1. Albert Camus, *Notebooks, 1935–1942* (New York: Modern Library, 1965), 130.

1. A Fighter Pilot's Entry into the Robotic Age

1. Immanuel Kant, "Formulae of the Categorical Imperative," in *Selections*, ed. Theodore Meyer Green (New York: Charles Scribner's Sons, 1929), 302.
2. Michael Walzer, *Just and Unjust Wars: A Moral Argument with Historical Illustrations* (New York: Basic Books, 2006), 43.
3. Michael Ignatieff, *The Warrior's Honor: Ethnic War and the Modern Conscience* (New York: Metropolitan Books, 1998), 161.
4. "Iraqi Miscalculation," *Time*, January 11, 1993, 10.
5. USAF fighters provided the preponderance of the air superiority assets and were primarily tasked with what was being called "defensive counter air" during both operations.
6. "World in Brief," *Washington Post*, February 2, 1999.
7. Common lore at what was then Joint Base Balad held that locals shelled the base in Saddam's time as well. During my time the base was hit ninety-eight times in five months, an experience not uncommon for anyone deployed to any operating location in Iraq. There were no direct injuries and only minor damage to facilities.
8. P. W. Singer, *Wired for War: The Robotics Revolution and Conflict in the Twenty-First Century* (New York: Penguin, 2009), 38.
9. Ibid., 38.
10. Armin Krishnan, *Killer Robots: Legality and Ethicality of Autonomous Weapons* (Burlington, VT: Ashgate, 2009), 4–5. Other definitions of robots will be discussed in a later section.
11. Viewing a weapon as a robot is a contentious viewpoint. Others argue a robot has to be able to alter its environment and then sense and act again as

part of a continuous loop. This only serves to show how little consensus there really is on what constitutes robotic technologies, as will be discussed later.

12. Carl H. Builder, *The Masks of War: American Military Styles in Strategy and Analysis* (Baltimore: Johns Hopkins University Press, 1989), 19.
13. Krishnan, *Killer Robots*, 33.

2. Now and When: Current and Future State of Robotics in Warfare

1. Available at "Prior National Defense Authorization Acts," last modified January 6, 2009, http://www.dod.mil/dodgc/olc/prior_ndaa.html (accessed January 23, 2013).
2. Singer, *Wired for War*, 59–60.
3. Ibid., 60.
4. Michael Ignatieff, *Virtual War: Kosovo and Beyond* (New York: Holt, 2000), 150. The RMA—the idea that precision ordnance so altered the way wars are fought that old concepts are obsolete—is still debated, but the argument is that precision-guided ordnance changes warfare by allowing direct targeting of military targets and creating immediate desired effects while protecting civilians and infrastructure, thereby lowering the risk to both combatants and noncombatants. It is often associated with the use of air power and stand-off weapons that further lower the risk to the combatants with an ability to use it.
5. Singer, *Wired for War*, 54.
6. Ibid., 22.
7. Ibid., 61.
8. Ibid.
9. Ibid.
10. P. W. Singer, "War of the Machines," *Scientific American* 303, no. 1 (July 2010): 56–63.
11. Ibid. As in the case of the iRobot PackBot in field testing in Afghanistan in 2001, urgent need requests went out for a technological solution to aid in countering improvised explosive devices. iRobot and other manufacturers rushed untested, non-program-of-record devices to soldiers in the field. Once their utility was proven or they saved the life of an explosive ordnance disposal technician, it fed the market for commercial off-the-shelf products to be fast-tracked into military use in the combat zone.
12. Comments during the official's lunch discussion with company grade officers at Joint Base Balad in 2008.
13. Department of Defense (hereafter DOD), *FY2009–2034 Unmanned Systems Integrated Roadmap* (Washington: DOD, 2009), 1. Hereafter referred to in the text as *Unmanned Systems Roadmap* or simply *Roadmap*.

14. Ibid., xiii–xiv.
15. Ibid., 10.
16. Department of the Air Force, *United States Air Force Unmanned Aircraft Systems Flight Plan, 2009–2047* (Washington: Department of the Air Force, 2009), 41.
17. DOD, *Roadmap*, 2.
18. Speaker to Senior Acquisition Course students, ICAF, November 2010.
19. DOD, *Roadmap*, 2–3.
20. Ibid., 3.
21. Ibid., 4.
22. Ibid.
23. Ibid., 5.
24. See DOD, DOD Directive Number 3000.09, *Autonomy in Weapon Systems* (Washington: Department of Defense, November 21, 2012), http://www.dtic.mil/whs/directives/corres/pdf/300009p.pdf.
25. DOD, *Roadmap*, 10.
26. Krishnan, *Killer Robots*, 33.
27. Department of the Air Force, *Unmanned Aircraft Systems Flight Plan*, 16.
28. Ibid.
29. Ibid., 41.
30. DOD, *Joint Robotics Program Master Plan FY2005* (Washington: Office of Under Secretary of Defense for Acquisition, Technology, and Logistics, 2005), 1–2.
31. Ibid., B-4.
32. Krishnan, *Killer Robots*, 4.
33. Singer, *Wired for War*, 67.
34. Ronald C. Arkin, and Lilia Moshkina, "Lethality and Autonomous Robots: An Ethical Stance," unpublished paper (Atlanta: College of Computing, Georgia Institute of Technology, 2007), http://www.cc.gatech.edu/ai/robot-lab/online-publications/ArkinMoshkinaISTAS.pdf (accessed September 26, 2010).
35. Discussions between ICAF faculty members and those of a university in the Boston area regarding support to ICAF's Robotics and Autonomous Systems Industry Study, Fall 2010.
36. DOD, *Joint Robotics Program*, B-1.
37. Krishnan, *Killer Robots*, 47.
38. DOD, *Joint Robotics Program*, B-1.
39. Many industrial robots now have vision sensors, are able to react to changes in their environment, and can compensate for those changes as they continue to properly perform their tasks.
40. DOD, *Joint Robotics Program*, B-1.

41. Krishnan, *Killer Robots*, 4.
42. Arkin and Moshkina, "Lethality and Autonomous Robots."
43. DOD, *Joint Robotics Program*, B-3.
44. Krishnan, *Killer Robots*, 43.
45. Comments of senior AUVSI corporate officers at an ICAF industry study seminar, February 2011.
46. International Society for Military Ethics conference, San Diego, January 25, 2011.
47. Missy Cummings, keynote address, International Society for Military Ethics conference, San Diego, January 25, 2011. More information and publications at http://web.mit.edu/aeroastro/labs/halab/index.shtml.
48. DOD, DOD Directive Number 3000.09, 14.
49. DOD, *Joint Robotics Program*, B-1. Currently this is the definition of "autonomy" in this document.
50. DOD Directive Number 3000.09 defines autonomous weapon systems as "a weapon system that, once activated, can select and engage targets without further intervention by a human operator. This includes human-supervised autonomous weapon systems that are designed to allow human operators to override operation of the weapon system, but can select and engage targets without further human input after activation."
51. Richard Hallion, *Storm over Iraq: Air Power and the Gulf War* (Washington: Smithsonian Institution Press, 1992), 171.
52. Ibid., 175.
53. Bernard E. Trainor, "Amphibious Operations in the Gulf War," *Marine Corps Gazette* 78, no. 8 (August 1994): 56.
54. Hallion, *Storm over Iraq*, 171.
55. Singer, *Wired for War*, 125.
56. These components are examples and are representative of surface-to-air missile systems. They are not all-inclusive and are not meant to represent any specific system.
57. The electromagnetic spectrum is not a recognized domain in U.S. military doctrine. Recognized domains are air, land, sea, space, and cyberspace. My contention is we have not categorized the cyber "domain" correctly. It operates wholly within the electromagnetic spectrum and ought to be considered just a sector of the electromagnetic domain.
58. Hallion, *Storm over Iraq*, 173.
59. Singer, *Wired for War*, 267.
60. DOD, *Roadmap*, 7.
61. B. Austen, "The Terminator Scenario," *Popular Science*, January 2011, 58.
62. DOD, *Roadmap*, 40.
63. This is an allusion to "Amara's Law," which will be discussed later.

3. The Law Is Not Enough

1. Henry Sidgwick, *The Elements of Politics* (London, New York: Macmillan, 1891), 253.
2. Christopher Coker, *Ethics and War in the 21st Century* (New York: Routledge, 2008), 148.
3. This occurred on November 19, 2010, during the question-and-answer session following a presentation to members of the Office of the Under Secretary of Defense for Policy by Ronald Arkin of the Georgia Institute of Technology on his work, under U.S. Army funding, on ethical autonomous lethality.
4. "Nonkinetic effects" is a term increasingly used to contrast some counterinsurgency operations with those involving the delivery of ordnance.
5. Walzer, *Just and Unjust Wars*, xxi.
6. Ibid., ix.
7. Ibid., 19.
8. Martin L. Cook, *The Moral Warrior: Ethics and Service in the U.S. Military* (Albany: State University of New York Press, 2004), 22.
9. Panel discussion at the AUVSI Program Review, February 1, 2011, Washington. AUVSI is the leading trade organization for unmanned and robotic vehicles and systems. More information at www.auvsi.org.
10. Leslie Claude Green, *The Contemporary Law of Armed Conflict* (Manchester and New York: Manchester University Press, 1993), 124.
11. Adam Roberts and Richard Guelff, eds., *Documents on the Laws of War*, 3rd ed. (New York: Oxford University Press, 2000), 8.
12. Ibid., 9.
13. Rupert Ticehurst, "The Martens Clause and the Laws of Armed Conflict," *International Review of the Red Cross* 37, no. 317 (April 30, 1997), http://www.icrc.org/eng/resources/documents/misc/57jnhy.htm (accessed November 29, 2010).
14. Green, *Contemporary Law of Armed Conflict*, 11.
15. Ibid.
16. Roberts and Guelff, *Documents on the Laws of War*, 646.
17. Ibid., 647.
18. Ibid., 170.
19. Ibid.
20. Ignatieff, *Virtual War: Kosovo and Beyond*, 199.
21. J. Storrs Hall, *Beyond AI: Creating the Conscience of the Machine* (Amherst, NY: Prometheus Books, 2007), 313.
22. Dave Grossman, *On Killing: The Psychological Cost of Learning to Kill in War and Society*, rev. ed. (New York: Little, Brown, 2009), 3.

23. Ibid., 23.
24. This is a descriptive "kill call," a way to remove aircraft from simulated battles in real time. We call the kind of aircraft, where it is, what altitude it is, and what it is doing at the time in order to remove the correct aircraft at the correct time. "No Name" is the name—ironic, I know—of a mountain on the Nellis Air Force Base tactics and training range in Nevada.
25. Civilian in the Office of the Secretary of Defense, correspondence with author, January 2011.
26. Walzer, *Just and Unjust Wars*, 142.
27. Ibid.
28. Ibid., 140.
29. Ibid., 142.
30. Ibid., 143.
31. Krishnan, *Killer Robots*, 115.
32. Michael Ignatieff, "Virtual War: Ethical Challenges," lecture, Center for the Study of Professional Military Ethics, U.S. Naval Academy, 2001, 17. Transcript available at http://www.usna.edu/ethics/publications/documents/IgnatieffPg1-24_Final.pdf.

4. The Spectra of Impunity in Warfare

1. Nikola Tesla, *My Inventions: The Autobiography of Nikola Tesla* (Radford, VA: Wilder Publications, 2007), 70.
2. Ibid., 77–78.
3. Micah Zenko, *Between Threats and War: U.S. Discrete Military Operations in the Post–Cold War World* (Stanford, CA: Stanford University Press, 2010), 73.
4. "U.S. Policy on Assassinations," CNN, November 4, 2002, http://archives.cnn.com/2002/LAW/11/04/us.assassination.policy/ (accessed January 30, 2011).
5. Martin van Creveld, *Technology and War: From 2000 B.C. to the Present* (New York: Free Press, 1991), 12.
6. Ibid., 12–13.
7. Green, *Contemporary Law of Armed Conflict*, 21.
8. Van Creveld, *Technology and War*, 71.
9. "Napoleon: Napoleon at War," Public Broadcasting System, 2000, http://www.pbs.org/empires/napoleon/n_war/campaign/page_12.html (accessed January 30, 2011).
10. Ibid., and Public Broadcasting System, "The Invasion of Iraq: Operation Iraqi Freedom," *Frontline*, February 26, 2004, http://www.pbs.org/wgbh/pages/frontline/shows/invasion/cron/ (accessed January 30, 2011).

11. Robert Coram, *Boyd: The Fighter Pilot Who Changed the Art of War* (Boston: Little, Brown, 2002), 334.
12. Thomas K. Adams, "Future Warfare and the Decline of Human Decision-making," *Parameters* 31, no. 4 (Winter 2001): 58.
13. Hallion, *Storm over Iraq*, 238.
14. "New Figures on Civilian Deaths in Kosovo War," Human Rights Watch, http://www.hrw.org/en/news/2000/02/07/new-figures-civilian-deaths-kosovo-war (accessed February 8, 2011).
15. Philip Alston, *Report of the Special Rapporteur on Extrajudicial, Summary or Arbitrary Executions* (New York: United Nations Human Rights Council, 2010), http://www2.ohchr.org/english/bodies/hrcouncil/docs/14session/A.HRC.14.24.Add6.pdf (accessed October 8, 2010).
16. David Kilcullen and Andrew McDonald, "Death from Above, Outrage Down Below," *New York Times*, May 17, 2009.
17. Alston, *Report of the Special Rapporteur*, 6.
18. They and others would no doubt question U.S. and Israeli intent to uphold these same principles, but there is, I believe, a difference. The United States and Israel are not engaged in random acts of violence with the aim of killing civilians as are the terrorists I've described here. There are valid questions to ask about legality, proportionality, and so forth, but there is a difference between these operations and purposely targeting civilians.
19. See Shaun R. McGrath, *Strategic Misstep: "Immortal" Robotic Warfare, Inviting Combat to Suburban America* (Carlisle, PA: U.S. Army War College, 2010), http://wwww.dtic.mil/cgi-bin/GetTRDoc?AD-ADA521822 (accessed June 23, 2010). David Whetham presented this idea at the International Society for Military Ethics conference in 2011 and in the book he edited, *Ethics, Law, and Military Operations* (New York: Palgrave Macmillan, 2010).
20. Hallion, *Storm over Iraq*, 284. See table 3 in appendix B for a comparison of World War II air strikes versus a theoretical strike in 1983.
21. Ibid., 199.
22. David A. Koplow, *Death by Moderation: The U.S. Military's Quest for Useable Weapons* (New York: Cambridge University Press, 2010), 67.
23. Michael Walzer, *Arguing about War* (New Haven, CT: Yale University Press, 2004), 14.
24. Ferenc Morton Szasz, *The Day the Sun Rose Twice: The Story of the Trinity Site Nuclear Explosion, July 16, 1945* (Albuquerque: University of New Mexico Press, 1984), 3.
25. Gar Alperovitz and Sanho Tree, *The Decision to Use the Atomic Bomb and the Architecture of an American Myth* (New York: Knopf, 1995), 847.
26. Cook, *Moral Warrior*, 124.

27. Brian Orend, "War," in *The Stanford Encyclopedia of Philosophy*, edited by Edward N. Zalta (Fall 2008), http://plato.stanford.edu/archives/fall2008/entries/war/ (accessed January 17, 2010).
28. Cook, *Moral Warrior*, 26.
29. Orend, "War."
30. Cook, *Moral Warrior*, 31.
31. Orend, "War."
32. Walzer, *Arguing about War*, 14.
33. Jeffrey Record, "Force-Protection Fetishism: Sources, Consequences, and (?) Solutions," *Air & Space Power Journal* 14, no. 2 (January 2000): 10.
34. Ignatieff, *Virtual War: Kosovo and Beyond*, 246. This is the thesis of this book where Ignatieff talks in depth on the concept of how killing with impunity is changing the face of warfare. Also note: Doctrinally there ought to be no such thing as an "air campaign" or a "land campaign." Operational plans in all domains make up the joint force commander's campaign; he or she is the only one who has a "campaign." The others are operations. However, in the case of Operation Allied Force, it truly was an air campaign because the sum total of the combatant commander's operational plan was executed from the air.
35. This is a common argument for unmanned aerial systems. It goes like this: since UASs are unmanned, they can fly at lower altitudes than manned systems, thereby increasing their sensor and weapon accuracy. This is only partially true, as other geometrical techniques can be used to increase accuracy. This misunderstanding of modern precision weaponry and the human actions required to properly employ it has a moderate to major effect on the discussion about the ethical and moral consequences, as evidenced by my observations at a conference of the International Society of Military Ethics in San Diego, January 25–28, 2011. Two things are at play: ethicists are quick to judge and explain technologies they have not made enough effort to understand, and (in this case) combat aviators have not been forthcoming enough about the technical skill and art of their craft. Improvement in both might add to the intellectual level of the debate.
36. Walzer, *Arguing about War*, 18.
37. These were the last two countries still willing to enforce the UN resolutions and demarches after Hussein successfully fractured world opinion during the cat-and-mouse game of the late 1990s and early 2000s over his suspected WMD program. It should serve as a word of warning to those who step into the breach when the international community makes such weighty decisions. Though Hussein never fully complied with the numerous resolutions in place and refused unfettered access to UN and International Atomic Energy Agency inspectors, he wore the world down over nine years until they all left it to the United Kingdom and the United States.

38. Larry May, Eric Rovie, and Steve Viner, eds., *The Morality of War: Classical and Contemporary Readings* (Upper Saddle River, NJ: Pearson Education, 2006), 160.
39. Walzer, *Just and Unjust Wars: A Moral Argument with Historical Illustrations*, 153.
40. Ibid.
41. Ibid.
42. Walzer, *Arguing about War*, 17.
43. Ibid., 18.
44. Ibid., 101.
45. Walzer, *Just and Unjust Wars*, xv.
46. "Drone Warfare," *CQ Researcher* 20, no. 28 (June 8, 2010): 662.
47. I will not wade into this discussion here. See Michael Gross's and Mary Ellen O'Connell's works cited in the bibliography, along with articles on drone operations in Afghanistan and Pakistan, for debates on this issue.
48. Charles J. Dunlap Jr., "Technology: Recomplicating Moral Life for the Nation's Defenders," *Parameters* 29, no. 3 (Autumn 1999): 27.
49. Michael L. Gross, *Moral Dilemmas of Modern War: Torture, Assassination, and Blackmail in an Age of Asymmetric Conflict* (New York: Cambridge University Press), 262.
50. Ibid., 30.
51. Walzer, *Just and Unjust Wars*, 36.
52. Ibid., 41.
53. Ibid., 36.
54. Ignatieff, *Virtual War: Kosovo and Beyond*, 161.
55. Christopher Coker, *Waging War without Warriors? The Changing Culture of Military Conflict* (Boulder, CO: Lynne Rienner, 2003), 31.
56. Mary Ellen O'Connell, "Respect the Battlefield," CBS News, http://www.cbsnews.com/stories/2010/04/08/opinion/main6377556.shtml (accessed October 20, 2010).
57. Coker, *Ethics and War*, 78. Citation is a memo from Attorney General Alberto Gonzales to President George W. Bush postulating the Geneva protocols are obsolete.
58. O'Connell, "Respect the Battlefield," 1.

5. Impunity and the Politics of War

1. John Ruskin, *The Crown of Wild Olive: Four Lectures on Industry and War* (New York: John Wiley & Sons, 1874), 92.
2. Walzer, *Just and Unjust Wars*, 265.

3. This was the declaration against Germany and Italy on December 11, 1941. On June 5, 1942, the United States also declared war on Romania, Hungary, and Bulgaria as they joined the Axis powers, with which U.S. forces were already in combat.
4. Koplow, *Death by Moderation*, 242.
5. Cook, *Moral Warrior*, 117.
6. Quote from *Spies like Us*, directed by John Landis (1985).
7. P. W. Singer, "The Ethics of Killer Applications: Why Is It So Hard to Talk about Morality When It Comes to New Military Technology?," *Journal of Military Ethics* 9, no. 4 (December 2010): 299–312.
8. None of this should be construed in any way as either support for or disagreement with Operation Iraqi Freedom. My opinions are personal and must remain so, as I explained to the woman at this convention.
9. Kenneth B. Moss, *Undeclared War and the Future of U.S. Foreign Policy* (Washington: Woodrow Wilson International Center for Scholars; Johns Hopkins University Press, 2008), 200.
10. Walzer, *Arguing about War*, 88.
11. Ibid., 89.
12. Walzer, *Just and Unjust Wars*, 29.
13. Hallion, *Storm over Iraq*, 195.
14. Ibid., 195–96.
15. See Mark Clodfelter's *The Limits of Air Power* (New York: The Free Press, 1989) for an aircentric point of view.
16. Christopher Cooper and John Harwood, "Americans Show Concern on Casualties: Nearly 40% of Those Polled Express Doubts That War Is Worth the Cost in U.S. Lives," *Wall Street Journal*, March 31, 2003.
17. Moss, *Undeclared War*, 206–7.
18. Cook, *Moral Warrior*, 80.
19. Ibid., 131.
20. Moss, *Undeclared War*, 200.
21. "Targeting Terror," *Current Events* 98, no. 1 (September 8, 1998): 3.
22. Koplow, *Death by Moderation*, 240.
23. Ibid.
24. Ibid., 241.
25. Cook, *Moral Warrior*, 92.
26. Dunlap, *Technology*, 41.
27. Eric Stoner, "Attack of the Killer Robots," in *Robotics*, ed. Kenneth Partridge (New York: H. W. Wilson, 2010), 111.
28. Ibid.
29. Steve Featherstone, "The Coming Robot Army," *Harper's Magazine*, February 2007, 52.

30. Mary Ellen O'Connell, interview by author, November 5, 2010, via phone from Burke, VA.
31. Alston, *Report of the Special Rapporteur*, 3.
32. Rafael Capurro and Michael Nagenborg, eds., *Ethics and Robotics* (Heidelberg, Germany: IOS Press, 2009), 84.
33. The Barack Obama administration changed the name of continuing operations. It is primarily a funding stream technicality and has little effect on overall operations.
34. Mary Ellen O'Connell, "Unlawful Killing with Combat Drones: A Case Study of Pakistan, 2004–2009," 3, http://ssrn.com/paper=1501144 (accessed February 20, 2011).
35. Ibid., 4.
36. Austen, *Terminator Scenario*, 58.
37. Rick Maze, "Draft Registration Should End, Lawmaker Says," *Army Times*, February 14, 2011, http://www.armytimes.com/news/2011/02/military-ending-draft-registration-021411w/ (accessed February 20, 2011).
38. Cook, *Moral Warrior*, ix.
39. David R. Segal and Mady Wechsler Segal, "America's Military Population," *Population Bulletin* 59, no. 4 (December 2004), http://www.prb.org/pdf04/59.4AmericanMilitary.pdf.
40. Senior Marine officer to students of National War College and ICAF, February 2011.
41. Camus, *Notebooks, 1935–1942*, 139.
42. Singer, *War of the Machines*, 56–63.
43. Ignatieff, *Virtual War: Kosovo and Beyond*, 181–84.
44. Ignatieff, "Virtual War: Ethical Challenges," 8.
45. Ignatieff, *Virtual War: Kosovo and Beyond*, 191.
46. Jutta Weber, "Robotic Warfare, Human Rights, and the Rhetorics of Ethical Machines," in Capurro and Nagenborg, *Ethics and Robotics*, 93.
47. Sidgwick, *Elements of Politics*, 254.
48. Coker, *Ethics and War*, 146.
49. Coker, *Waging War without Warriors?*, 71.
50. Stoner, "Attack of the Killer Robots," in Partridge, *Robotics*, 113.
51. DARPA independently contracted engineer, remarks to the Robotics and Autonomous Systems Industry Study, ICAF, January 2011.
52. Singer, *Wired for War*, 22.
53. DARPA independently contracted engineer, remarks to the Robotics and Autonomous Systems Industry Study.
54. Ibid.
55. Cook, *Moral Warrior*, 89–90.

56. Colin L. Powell and Joseph E. Persico, *My American Journey* (New York: Random House, 1995), 576.
57. Stephen Wrage, "When War Isn't Hell: A Cautionary Tale," *Current History* 102, no. 660 (January 2003): 32.
58. Harald Müller, "The Antinomy of Democratic Peace," *International Politics* 41, no. 4 (December 2004): 500.
59. Coker, *Waging War without Warriors?*, 63.
60. Dunlap, *Technology*, 35.
61. Koplow, *Death by Moderation*, 47.

6. Impunity and the Warrior

1. Camus, *Notebooks, 1935–1942*, 149.
2. Featherstone, "Coming Robot Army," 46.
3. Coker, *Waging War without Warriors?*, 54.
4. Ibid., 33.
5. This is not exclusive to RPA operators. Some bomber and fighter crews flying in Operation Allied Force also experienced this. Those assigned to Aviano Air Base flew combat from their home base in Italy. B-2 pilots flew combat sorties from Whiteman Air Force Base in Missouri, where they were stationed and where their families lived.
6. Singer, *Wired for War*, 253.
7. Ibid., 252.
8. DOD, *Roadmap*, 40.
9. Singer, *Wired for War*, 252.
10. Singer, *War of the Machines*, 56–63.
11. Singer, *Wired for War*, 253.
12. George Leonard, "Introduction," *In Search of the Warrior Spirit*, ed. Richard Strozzi-Heckler (Berkeley, CA: North Atlantic Books, 1990), xiii.
13. Ibid., 6.
14. Coker, *Waging War without Warriors?*, 58.
15. Leonard, "Introduction," xi.
16. Coker, *Waging War without Warriors?*, 67.
17. Strozzi-Heckler, *In Search of the Warrior Spirit*, 66.
18. Singer, *Wired for War*, 331.
19. The term "combat air forces" is used in the USAF to describe aircraft whose primary function is direct combat action. Fighters, bombers, some electronic attack aircraft, etc., are considered part of the CAF.
20. Mary Ellen O'Connell, "Flying Blind," *America* 202, no. 8 (March 15, 2010): 13.
21. Coker, *Waging War without Warriors?*, 123.

22. Walzer, *Arguing about War*, 16.
23. Ibid., 16.
24. Coker, *Waging War without Warriors?*, 59.
25. The letter *g* is the symbol for the acceleration of gravity. In a dogfight pilots undergo many times the force of gravity in order to make their aircraft turn tighter, faster, and for longer than their opponent.
26. Strozzi-Heckler, *In Search of the Warrior Spirit*, 62.
27. Grossman, *On Killing*, 4.
28. Partridge, *Robotics*, 110.
29. Coker, *Waging War without Warriors?*, 173.
30. Krishnan, *Killer Robots*, 128.
31. Weber, "Robotic Warfare," in Capurro and Nagenborg, *Ethics and Robotics*, 91.
32. Coker, *Waging War without Warriors?*, 176.
33. Partridge, *Robotics*, 110.
34. Grossman, *On Killing*, 100.
35. Koplow, *Death by Moderation*, 42. Consider the case of the commuter train during Operation Allied Force that entered the weapon system operator's (the aircrew member guiding the bomb) view seconds before it impacted and destroyed the bridge the train was about to transit. Innocents were killed, and the public reaction was significant.
36. The USAF Weapons School is the Air Force's premier tactical flying school, where the course of instruction is a grueling six months of flying and academics. We often refer to it as a "doctorate in tactical aviation."
37. O'Connell, *Flying Blind*, 13.
38. Fellow instructor at the USAF Weapons School, conversation with author, circa 2002.
39. O'Connell, *Flying Blind*, 13.
40. This is certainly a stretch, and it would be unexpected for them to don a uniform and/or carry arms openly. However, while we might find cause to prosecute them for violations of laws of war on such grounds, we could not argue in any meaningful way that their ultimate ends were not justified.
41. Dunlap, *Technology*, 30.
42. This scenario is borrowed from a talk presented by David Whetham, professor at King's College London and the Joint Services Command and Staff College, to the International Society for Military Ethics conference, San Diego University, January 2011.
43. A common refrain of civilian authorities in our "persistent conflict."
44. Partridge, *Robotics*, 111.
45. William W. Taylor, J. H. Bigelow and John A. Ausink, *Fighter Drawdown Dynamics: Effects on Aircrew Inventories* (Santa Monica, CA: Rand Corp., 2009),

68, http://www.rand.org/pubs/monographs/2009/RAND%5FMG855.pdf (accessed February 23, 2011).

46. Ibid., 53.
47. Ibid., 89.
48. Unpublished proceedings of TAMI 21 studies on staff requirements for fighter aviators.
49. There are exceptions. Military personnel seem more concerned than others, which gives some hope.
50. DARPA SETA, discussion with author, AUVSI Program Review, Washington, February 2011. Paraphrased.
51. Defense Advanced Research Projects Agency website, http://www.darpa.mil/About.aspx (accessed March 2, 2011).
52. Singer, *Wired for War*, 173.
53. Thomas H. Cowan Jr., *Theoretical, Legal and Ethical Impact of Robots on Warfare* (Carlisle, PA: U.S. Army War College, 2007), 7.
54. Author knowledgeable on robotics and war during comments to Senior Acquisition Concentration seminar, ICAF, November 2010.
55. Former CAOC director, Operation Iraqi Freedom, in comments to students of the 2005 academic year, Air Command and Staff College, Montgomery, AL.
56. Department of the Air Force, *Air Force Doctrine Document 1, Air Force Basic Doctrine*, Headquarters, Air Force Doctrine Center, Maxwell Air Force Base, AL (Washington: Department of the Air Force, 2003), 28.
57. P. W. Singer, "Robots and the Rise of 'Tactical Generals,'" *Defense News*, March 9, 2009, 29–29.
58. The general in question, and many others I'm sure, would dispute my supposition. I asked him once why he did it and whether the Air Force should just give up and start embracing "centralized execution" in doctrine. His rationale for taking such direct control was to take the responsibility for any collateral damage on his own shoulders versus those of the young airman operating the sensor. He would argue he was concerned with strategic effects—collateral damage gone bad—in this instance, and he would have a very good point.
59. Singer, "Robots and the Rise."
60. Ibid.
61. Coker, *Ethics and War*, 61.
62. Krishnan, *Killer Robots*, 136.
63. Department of the Air Force, *Unmanned Aircraft Systems Flight Plan*, 51.
64. DOD, *Unmanned Systems Roadmap*, 39.

65. Author impression from various UAS, UGV, and UMV panels as part of the ICAF Robotics and Autonomous Systems Industry Study Group, spring 2011.
66. Adams, *Future Warfare*, 60.
67. Ron Arkin highlighted this nature at a discussion in the Office of the Under Secretary of Defense for Policy (OSD Policy), Arlington, VA, November 2010.
68. Krishnan, *Killer Robots*, 136.
69. Ibid.
70. Ibid.
71. TAMI 21 proposed numerous fixes to solve fighter pilot manning problems. "Requirements" for experienced fighter aviators was just one part of a holistic plan. For more information, see Taylor et al., *Fighter Drawdown Effects*.
72. Ronald Arkin, *Governing Lethal Behavior in Autonomous Robots* (Boca Raton, FL: CRC Press, 2009), 146.
73. DOD, *Unmanned Systems Roadmap*, 40.
74. Dunlap, *Technology*, 43–44.

7. Impunity and the Future of War

1. Zenko, *Between Threats and War*, 90.
2. "History: Products; North American Aviation OV-10 Bronco," Boeing Company, http://www.boeing.com/history/bna/ov10.htm (accessed March 4, 2011).
3. Stephen Trimble, "Boeing Considers Restarting OV-10 Production after 23-Year Hiatus," Flightglobal, February 1, 2009, http://www.flightglobal.com/articles/2009/02/01/321730/boeing-considers-restarting-ov-10-production-after-23-year.html (accessed March 4, 2011).
4. Thom Shanker, "Gates Warns against Wars like Iraq and Afghanistan," *New York Times*, February 26, 2011.
5. George W. Casey Jr., "America's Army in an Era of Persistent Conflict," *Army* 58, no. 10 (October 2008): 19.
6. Rupert Smith, *The Utility of Force: The Art of War in the Modern World* (New York: Vintage, 2008), 271.
7. Ibid., 269.
8. Ibid., 3.
9. Ibid., 271.
10. Ibid.
11. Ibid.

12. James F. Amos and David H. Petraeus, *Field Manual (FM) 3-24, Counterinsurgency* (Washington: Department of the Army, 2006), 3–13, http://www.fas.org/irp/doddir/army/fm3-24.pdf (accessed March 8, 2011).
13. Ibid., A-5.
14. George Casey Jr., "The Second Decade," *Army* 60, no. 10 (October 2010): 19.
15. Krishnan, *Killer Robots*, 35.
16. DOD, *Unmanned Systems Roadmap*, 39.
17. Members of the RS JPO and this facility do not like to call it a "depot." It is not one of the DOD's official depots, but it essentially functions as one.
18. "Remarks by Secretary Gates at the United States Air Force Academy." *U.S. Department of Defense Public Affairs Release* (March 4, 2011). http://www.defense.gov/transcripts/transcript.aspx?transcriptid=4779 (accessed March 10, 2011).
19. Coker, *Waging War without Warriors?*, 95.
20. Smith, *Utility of Force*, 3.
21. Martin Cook, comments at the 2011 International Society for Military Ethics conference, San Diego, January 2011. Also discussed in his book *The Moral Warrior.*
22. "Iraqi Aircraft 'Buried in Desert,'" BBC, http://news.bbc.co.uk/2/hi/middle_east/3116259.stm (accessed March 5, 2011).
23. RS JPO briefing, Detroit, MI, February 2011.
24. Capurro and Nagenborg, *Ethics and Robotics*, 84.
25. Koplow, *Death by Moderation*, 247.
26. Coker, *Waging War without Warriors?*, 62.
27. Koplow, *Death by Moderation*, 246.
28. Gross, *Moral Dilemmas*, 119.
29. Ibid.
30. Koplow, *Death by Moderation*, 246.
31. Gross, *Moral Dilemmas*, 239.
32. Ibid.
33. Ibid., 263.
34. Ignatieff, *Virtual War: Kosovo and Beyond*, 150–51.
35. Koplow, *Death by Moderation*, 41.
36. Singer, *Ethics of Killer Applications*, 310.
37. Ibid., 306.
38. Coker, *Waging War without Warriors?*, 75.
39. Ibid., 82.
40. Ibid., 160.
41. Krishnan, *Killer Robots*, 127.
42. Coker, *Ethics and War*, 61.
43. Gross, *Moral Dilemmas*, 238–39.

8. AI, the Search for Relevance, and Robotic *Jus in Bello*

1. Arkin, *Governing Lethal Behavior*, 29.
2. Capurro and Nagenborg, *Ethics and Robotics*, 95.
3. Singer, *Wired for War*, 104.
4. Hall, *Beyond AI*, 265.
5. J. Storrs Hall, "Ethics for Machines," Autogeny.org, http://autogeny.org/ethics.html (accessed November 5, 2010).
6. Lincoln Barnett, "J. Robert Oppenheimer," *Life*, October 10, 1949, 133, http://books.google.com/books?id=GVIEAAAAMBAJ&pg=PA120&lpg=PA120&dq=Robert+Oppenheimer+life+magazine+1949&source=bl&ots=ra0Wvofoms&sig=JBeYL42CFgxLfOoYgyYaThPrO30&hl=en&ei=MXN9Tf-WOaHE0QG_t5DPAw&sa=X&oi=book_result&ct=result&resnum=3&ved=0CCQQ6AEwAg#v=onepage&q&f=false (accessed March 13, 2011).
7. Robert Cahn, "The Day the World Changed, July 16, 1945. Part III: Elation Gives Place to Contemplation," *Christian Science Monitor* (Pre-1997 Fulltext), July 13, 1995. See also http://www.youtube.com/watch?v=f94j9WIWPQQ for a portion of an interview conducted in 1965 and aired on NBC (accessed March 9, 2011).
8. Patrick Lin, George Bekey, and Keith Abney, "Robots in War: Issues of Risk and Ethics," in Capurro and Nagenborg, *Ethics and Robotics*, 52.
9. Cook, *Moral Warrior*, 33.
10. Roberts and Guelff, *Documents on the Laws of War*, 10.
11. Walzer, *Just and Unjust Wars*, 287.
12. Ibid., 288.
13. Hall, *Beyond AI*, 62.
14. Ibid., 147.
15. Ibid., 63. Hall is describing the question Turing posed. The term "teletype" has come to be understood as a textual interface.
16. Ibid., 64.
17. Ibid., 147.
18. Clark Glymour, "Android Epistemology and the Frame Problem: Comments on Dennet's 'Cognitive Wheels,'" in *The Robot's Dilemma: The Frame Problem in Artificial Intelligence*, ed. Zenon W. Pylyshyn (Norwood, NJ: Ablex, 1987), 67.
19. Daniel C. Dennett, "Cognitive Wheels: The Frame Problem of AI," in Pylyshyn, *Robot's Dilemma*, 42, 63. McCarthy and Hayes's original paper can be found here: http://www-formal.stanford.edu/jmc/mcchay69.pdf.
20. Glymour, "Android Epistemology," in Pylyshyn, *Robot's Dilemma*, 65.
21. Patrick J. Hayes, "What the Frame Problem Is and Isn't," in Pylyshyn, *Robot's Dilemma*, 125.

22. Ibid.
23. Ibid.
24. Ibid.
25. Dennett, "Cognitive Wheels," in Pylyshyn, *Robot's Dilemma*, 41–42.
26. Industrial Light and Magic is the name of George Lucas's production company, the makers of the famous Star Wars films.
27. Drew McDermott, "We've Been Framed: Or, Why AI Is Innocent of the Frame Problem," in Pylyshyn, *Robot's Dilemma*, 116.
28. Hayes, "What the Frame Problem," 124.
29. In the example that follows, the broad categories of ROE discussed are only possibilities of what may appear in various ROE. They are not representative of any current or active ROE or any particular operational plan.
30. Arkin, *Governing Lethal Behavior*, xvi.
31. Ibid., 29–30.
32. Comments to meeting in OSD Policy, Arlington, VA, November 19, 2010.
33. Featherstone, "Coming Robot Army," 49.
34. Arkin, *Governing Lethal Behavior*, 96.
35. Ronald Arkin, phone interview with the author, November 12, 2010, Burke, VA.
36. Arkin, *Governing Lethal Behavior*, 120.
37. Ibid., 146.
38. Arkin, interview with the author.
39. Arkin, *Governing Lethal Behavior*, 104. Arkin cites the final report of Mental Health Advisory Team IV (under the auspices of the Office of the Surgeon, Multinational Force–Iraq, and the Office of the Surgeon General, U.S. Army Medical Command), dated November 17, 2006, for concerns over current combatants' mind-set on the laws of war and treatment of civilians and combatants. It is available at http://www.armymedicine.army.mil/reports/mhat/mhat_iv/mhat-iv.cfm.
40. Arkin, interview with the author.
41. Unpublished briefing to ICAF's Robotics and Autonomous Systems Industry Study, Fort McNair, Washington, D.C., March 8, 2011.
42. Ibid.
43. Ibid.
44. Ibid.
45. Krishnan, *Killer Robots*, 98.
46. Ibid., 99.
47. Ibid., 109.
48. ONR researcher comments to ICAF's Robotics and Autonomous Systems Industry Study, Fort McNair, Washington, D.C., March 8, 2011.
49. Krishnan, *Killer Robots*, 109.

50. Inspired by a story by Arthur C. Clarke, *2001: A Space Odyssey* was produced and directed by Stanley Kubrick.
51. Daniel C. Dennett, "When HAL Kills, Who's to Blame?," in *HAL's Legacy: 2001's Computer as Dream and Reality*, ed. David G. Stork (Cambridge, MA: MIT Press, 1997).
52. Weber, "Robotic Warfare," in Capurro and Nagenborg, *Ethics and Robotics*, 95. Weber is quoting computer scientist Noel Sharkey.
53. Dennett, "When HAL Kills," 354.
54. Arkin, *Governing Lethal Behavior*, 148.
55. Arkin, interview with the author.
56. Featherstone, "Coming Robot Army," 49.
57. Hall, *Beyond AI*, 269.
58. Ibid., 269.
59. Ibid., 270.
60. Ibid., 271.
61. Ibid.
62. Ibid.
63. Ibid., 273.
64. Krishnan, *Killer Robots*, 103.
65. Jerry Adler Jr. with Frank Gibney, "After the Challenger," *Newsweek*, October 10, 1988, 28.
66. Weber, "Robotic Warfare," in Capurro and Nagenborg, *Ethics and Robotics*, 92.
67. It may be surprising to know programmers and engineers do not always know exactly how a system will perform, but this is really the basis of experimentation and testing. We design systems and see if they do what we think they will do. In the grand scheme they often do, but invariably engineers will find their machines reacting in ways they did not expect. This is what leads to better and better products. The problem is as machines get more and more complex, finding ways to test extensively gets harder and more expensive. There is no way to test absolutely everything, and therefore some systems that make it to market still have flaws. You know this implicitly if your car has ever been recalled. See DOD Directive Number 3000.09 for the official DOD stance on test and evaluation of autonomous and semi-autonomous weapon systems.
68. Weber, "Robotic Warfare," in Capurro and Nagenborg, *Ethics and Robotics*, 92.
69. Featherstone, "Coming Robot Army," 49.
70. Discussions with the author, ICAF, Fort McNair, Washington, D.C., March 8, 2011.
71. However, I contend it still does nothing to solve the internal culture issues or problems with moral equality of the operator and other combatants.

72. Lin, Bekey, and Abney, *Autonomous Military Robots*, 64.
73. Ibid.
74. Weber, "Robotic Warfare," in Capurro and Nagenborg, *Ethics and Robotics*, 93.
75. Barnett, "J. Robert Oppenheimer," 136.

9. Radical Responsibilities

1. Walzer, *Just and Unjust Wars*, 226.
2. Ibid.
3. Ibid., 225–26.
4. Conversation with the author, March 2011.
5. Walzer, *Just and Unjust Wars*, 227.
6. Ibid.
7. Ibid.
8. Anne Kornblut, "In Interview, Bush Defends Iraq War and Waterboarding," *Washington Post*, November 9, 2010, http://www.washingtonpost.com/wp-dyn/content/article/2010/11/08/AR2010110807347.html (accessed March 15, 2011).
9. One hopes they would not think it either, and this makes President Bush's statement even more astounding. He also said in the interview the lawyers told him it was legal and that was the impetus to use these techniques. We have talked before about the difference between legal and right, and the president's verbiage in the quote above highlights the slightly different context.
10. Based on Internet-sourced photos and remarks by Ronald Arkin during a briefing to OSD Policy, Arlington, VA, November 19, 2010. I have no personal experience with ROE in Operation Enduring Freedom, and I did not review it for this unclassified research. I am making no reference to actual ROE.
11. Coker, *Ethics and War*, 6.
12. I borrow and expand on a term coined by Michael Walzer in his book *Arguing about War.* Specific citations follow.
13. Walzer, *Arguing about War*, 14.
14. Ibid., 24.
15. Ibid., 25.
16. Ibid., 24.
17. Ibid., 25.
18. Ibid., 28.
19. Ibid. This paragraph summarizes the argument on pages 24–28.
20. Recall proportionality is twice required.

21. Every time there is a terrorist threat on board a commercial passenger jet, there is the possibility of that aircraft being shot down by U.S. fighters, but this is slightly different. Those decisions would most likely be made under a purely utilitarian argument. If, for instance the aircraft was hijacked and headed for a nuclear facility, the greater good would be to shoot it down before it arrived, hoping for it to fall in an empty field. The lives lost on board the aircraft, while tragic, would be outweighed by the lives saved. Importantly they are all innocent lives being weighed—those on the ground and those in the air. Additionally, there is likely an assumption the lives on board the aircraft would be lost no matter what military action or inaction takes place. Justified killing of noncombatants is not quite the same. First, their deaths do not seem inevitable, but more importantly their lives are not weighed against the loss of other innocent life. Their lives are weighed against the legitimacy of the act of war, its direct moral effect, the intention of the killer, and only then the utilitarian argument that the good outweighs the evil effect of killing innocents. In the world of the risk inversion, these kinds of calculations may be taking place on the streets of suburbia as authorities, perhaps military members, root out combatants from within our own population.
22. Walzer, *Arguing about War*, 32.
23. *National Security Strategy: May 2010* (Washington: Office of the President of the United States, 2010), 1, http://www.whitehouse.gov/sites/default/files/rss_viewer/national_security_strategy.pdf (accessed March 16, 2011).
24. "World Wide Military Expenditures—2011," Global Security, http://www.globalsecurity.org/military/world/spending.htm (accessed March 16, 2011).
25. Nat Hentoff, "Few Batting Eyes at Obama's Deadly Drone Policy," *The Trentonian*, July 28, 2010, http://www.trentonian.com/articles/2010/07/28/opinion/doc4c50f55db659a777910565.txt (accessed October 20, 2010).
26. Walzer, *Just and Unjust Wars*, 262.
27. "Frag" is a term used to describe the effective radius of explosives in doing their designed work. It is short for "fragmentation."
28. Walzer, *Just and Unjust Wars*, 16.
29. Coker, *Ethics and War*, 13.
30. Gross, *Moral Dilemmas*, 231.
31. Ibid., 235.
32. Ibid., 238.
33. Ibid.
34. Walzer, *Just and Unjust Wars*, 143.
35. Cook, *Moral Warrior*, 117.
36. Ignatieff, *Virtual War: Kosovo and Beyond*, 198.
37. Ibid., 104.

38. Mental Health Advisory Team IV, *Final Report*, 36.
39. Ibid.
40. Ibid.
41. Ibid., 35.
42. Ibid.
43. Charlie Savage, "Case of Soldiers Accused in Afghan Civilian Killings May Be Worst of Two Wars," *New York Times*, October 4, 2010.
44. Senior general officer assigned to the International Security and Assistance Forces, Afghanistan, National Defense University, March 2011.
45. Cook, *Moral Warrior*, 40.
46. Coker, *Ethics and War*, 15.
47. H. Norman Schwarzkopf and Peter Petre, *It Doesn't Take a Hero: General H. Norman Schwarzkopf, the Autobiography* (New York: Bantam, 1993), 542.
48. Ibid.
49. Ibid., 543.
50. Ibid., 542.
51. Walzer, *Arguing about War*, 97.
52. Ignatieff, *Virtual War: Kosovo and Beyond*, 161.
53. Walzer, *Arguing about War*, 97.
54. Coker, *Ethics and War*, 151.
55. Kant, *Selections*, 302.

10. Inevitability, Persistence . . . and Heart

1. Omar N. Bradley, "Armistice Day 1948 Address: General Omar N. Bradley," Opinionbug.com, http://www.opinionbug.com/2109/armistice-day-1948-address-general-omar-n-bradley/ (accessed February 19, 2011).
2. Walzer, *Just and Unjust Wars*, 265.
3. Coker, *Waging War without Warriors?*, 81.
4. Coker, *Ethics and War*, 150.
5. Dion Hinchcliffe, "Twenty-Two Power Laws of the Emerging Social Economy," ZDNet, October 5, 2009, http://www.zdnet.com/blog/hinchcliffe/twenty-two-power-laws-of-the-emerging-social-economy/961 (accessed March 20, 2011).
6. Comments to the AUVSI Program Review, Washington, D.C., February 2, 2011.
7. Results of Robotics and Autonomous Systems Industry Study business strategy assessment of four firms and discussions on site with another.
8. Dan Ephron and Kevin Peraino, "Hizbullah's Worrisome Weapon," *Newsweek*, September 11, 2006, 28–28.

9. This idea was debated at the annual meeting of the International Society for Military Ethics Conference in San Diego, in January 2011. The author specifically discussed this idea with George Lucas, professor of ethics and public policy at the Naval Postgraduate School in Monterey, California, and chair in ethics, Stockdale Center for Ethical Leadership, U.S. Naval Academy, in relation to a proposal for a similar council proposed to DARPA.
10. Discussion with a senior executive for an unmanned systems trade organization, ICAF, Fort McNair, Washington, D.C., January 2011.
11. American Battle Monuments Commission, http://www.abmc.gov/cemeteries/cemeteries/no.php (accessed March 29, 2011).
12. Walzer, *Just and Unjust Wars*, 265.
13. Ibid., 32.
14. Martin van Creveld, *The Transformation of War* (New York: Collier Macmillan, 1991), 226.
15. Ignatieff, *Warrior's Honor*, 118.
16. Dunlap, *Technology*, 34–35.
17. Coker, *Ethics and War*, 149–50.
18. Ignatieff, *Warrior's Honor*, 161.
19. Coker, *Ethics and War*, 24.

SELECTED BIBLIOGRAPHY

Adams, Thomas K. "Future Warfare and the Decline of Human Decisionmaking." *Parameters* 31, no. 4 (Winter 2001): 57.

Adler, Jerry, Jr., with Frank Gibney. "After the Challenger." *Newsweek*, October 10, 1988, 28.

Alperovitz, Gar, and Sanho Tree. *The Decision to Use the Atomic Bomb and the Architecture of an American Myth*. New York: Knopf, 1995.

Alston, Philip. *Report of the Special Rapporteur on Extrajudicial, Summary or Arbitrary Executions*. New York: United Nations Human Rights Council, 2010.

Amos, James F., and David H. Petraeus. *Field Manual (FM) 3-24, Counterinsurgency*. Washington: Department of the Army, 2006.

Arkin, Ronald. *Governing Lethal Behavior in Autonomous Robots*. Boca Raton, FL: CRC Press, 2009.

Arkin, Ronald C., and Lilia Moshkina. "Lethality and Autonomous Robots: An Ethical Stance." Unpublished paper. Atlanta: College of Computing, Georgia Institute of Technology. http://www.cc.gatech.edu/ai/robot-lab/online-publications/ArkinMoshkinaISTAS.pdf (accessed September 26, 2010).

Austen, B. "The Terminator Scenario." *Popular Science*, January 2011, 58.

Barnett, Lincoln. "J. Robert Oppenheimer." *Life*, October 10, 1949, 120.

Boeing Company. "History: OV-10 Bronco." http://www.boeing.com/history/bna/ov10.htm (accessed March 4, 2011).

Bradley, Omar N. "Armistice Day 1948 Address: General Omar N. Bradley." Opinionbug.com. http://www.opinionbug.com/2109/armistice-day-1948-address-general-omar-n-bradley/ (accessed February 19, 2011).

British Broadcasting Corporation. "Iraqi Aircraft 'Buried in Desert,'" August 1, 2003. http://news.bbc.co.uk/2/hi/middle_east/3116259.stm (accessed March 5, 2011).

Builder, Carl H. *The Masks of War: American Military Styles in Strategy and Analysis*. Rand Corp. research study. Baltimore: Johns Hopkins University Press, 1989.

Cahn, Robert. "The Day the World Changed, July 16, 1945. Part III: Elation Gives Place to Contemplation." *Christian Science Monitor*, July 13, 1995.

Camus, Albert. *Notebooks, 1935–1942*. New York: Modern Library, 1965.

Capurro, Rafael, and Michael Nagenborg, eds. *Ethics and Robotics*. Heidelberg, Germany: IOS Press, 2009.

Casey, George W., Jr. "America's Army in an Era of Persistent Conflict." *Army* 58, no. 10 (October 2008): 19.

———. "The Second Decade." *Army* 60, no. 10 (October 2010): 19.

CNN. "U.S. Policy on Assassinations," November 4, 2002. http://archives.cnn.com/2002/LAW/11/04/us.assassination.policy/ (accessed January 30, 2011).

Coker, Christopher. *Ethics and War in the 21st Century*. New York: Routledge, 2008.

———. *Waging War without Warriors? The Changing Culture of Military Conflict*. Boulder, CO: Lynne Rienner, 2002.

Cook, Martin L. *The Moral Warrior: Ethics and Service in the U.S. Military*. Albany: State University of New York Press, 2004.

Cooper, Christopher, and John Harwood. "Americans Show Concern on Casualties: Nearly 40% of Those Polled Express Doubts That War Is Worth the Cost in U.S. Lives." *Wall Street Journal*, March 31, 2003.

Coram, Robert. *Boyd: The Fighter Pilot Who Changed the Art of War*. Boston: Little, Brown, 2002.

Cowan, Thomas H., Jr. *Theoretical, Legal and Ethical Impact of Robots on Warfare*. Carlisle, PA: U.S. Army War College, 2007.

CQ Researcher. "Drone Warfare." 20, no. 28 (August 2010): 653–76.

Defense Advanced Research Projects Agency.

Dennett, Daniel C. "When HAL Kills, Who's to Blame?" In *HAL's Legacy: 2001's Computer as Dream and Reality*, David G. Stork, ed. Cambridge, MA: MIT Press, 1997.

Department of the Air Force. *Air Force Doctrine Document 1, Air Force Basic Doctrine*. Edited by Headquarters, Air Force Doctrine Center, Maxwell Air Force Base, AL. Washington: Department of the Air Force, 2003.

———. *United States Air Force Unmanned Aircraft Systems Flight Plan 2009–2047*. Washington: Department of the Air Force, 2009.

Department of Defense. *Department of Defense Directive Number 3000.09, Autonomy in Weapon Systems.* Washington: Department of Defense, 2012.

———. *Department of Defense Directive Number 7045.20, Capability Portfolio Management.* Washington: Department of Defense, 2008.

———. *FY2009–2034 Unmanned Systems Integrated Roadmap*. Washington: Department of Defense, 2009.

———. *Joint Robotics Program Master Plan FY2005.* Washington: Office of Under Secretary of Defense for Acquisition, Technology, and Logistics, 2005.

Dunlap, Charles J., Jr. "Technology: Recomplicating Moral Life for the Nation's Defenders." *Parameters* 29, no. 3 (Autumn 1999): 24.

Ephron, Dan, and Kevin Peraino. "Hizbullah's Worrisome Weapon." *Newsweek*, September 11, 2006.

Featherstone, Steve. "The Coming Robot Army." *Harper's Magazine*, February 2007, 43–52.

Gasser, Hans-Peter. "The U.S. Decision Not to Ratify Protocol I to the Geneva Conventions on the Protection of War Victims: An Appeal for Ratification by the United States," *American Journal of International Law* 74 (October 1987).

Global Security. "World Wide Military Expenditures—2011." http://www.globalsecurity.org/military/world/spending.htm (accessed March 16, 2011).

Green, Leslie Claude. *The Contemporary Law of Armed Conflict*. New York: Manchester University Press, 1993.

Gross, Michael L. *Moral Dilemmas of Modern War: Torture, Assassination, and Blackmail in an Age of Asymmetric Conflict.* New York: Cambridge University Press, 2010.

Grossman, Dave. *On Killing: The Psychological Cost of Learning to Kill in War and Society*. Rev. ed. New York: Little, Brown, 2009.

Hall, J. Storrs. *Beyond AI: Creating the Conscience of the Machine*. Amherst, NY: Prometheus Books, 2007.

———. "Ethics for Machines." Autogeny.org, 2000. http://autogeny.org/ethics.html (accessed November 5, 2010).

Hallion, Richard. *Storm over Iraq: Air Power and the Gulf War*. Washington: Smithsonian Institution Press, 1992.

Hentoff, Nat. "Few Batting Eyes at Obama's Deadly Drone Policy." *The Trentonian*, July 28, 2010. http://www.trentonian.com/articles/2010/07/28/opinion/doc4c50f55db659a777910565.txt (accessed October 20, 2010).

Hinchcliffe, Dion. "Twenty-Two Power Laws of the Emerging Social Economy." ZD Net, October 5, 2009. http://www.zdnet.com/blog/hinchcliffe/twenty-two-power-laws-of-the-emerging-social-economy/961 (accessed March 20, 2011).

Human Rights Watch. "New Figures on Civilian Deaths in Kosovo War," February 8, 2000. http://www.hrw.org/en/news/2000/02/07/new-figures-civilian-deaths-kosovo-war (accessed February 8, 2011).

Ignatieff, Michael. "Virtual War: Ethical Challenges." Abridged transcript from third lecture in series sponsored by the Center for the Study of Professional Military Ethics. Annapolis, MD: Center for the Study of Professional Military Ethics, United States Naval Academy, 2001.

———. *Virtual War: Kosovo and Beyond.* New York: Holt, 2000.

———. *The Warrior's Honor: Ethnic War and the Modern Conscience.* New York: Metropolitan Books, 1998.

Kant, Immanuel. *Selections.* Theodore Meyer Green, ed. New York: Charles Scribner's Sons, 1929.

Kilcullen, David, and Andrew McDonald. "Death from Above, Outrage Down Below." *New York Times*, May 17, 2009.

Koplow, David A. *Death by Moderation: The U.S. Military's Quest for Useable Weapons.* New York: Cambridge University Press, 2010.

Kornblut, Anne. "In Interview, Bush Defends Iraq War and Waterboarding." *Washington Post,* November 9, 2010. http://www.washingtonpost.com/wp-dyn/content/article/2010/11/08/AR2010110807347.html (accessed March 15, 2011).

Krishnan, Armin. *Killer Robots: Legality and Ethicality of Autonomous Weapons.* Burlington, VT: Ashgate, 2009.

Lin, Patrick, George Bekey, and Keith Abney. *Autonomous Military Robots: Risk, Ethics, and Design.* San Luis Obispo: California Polytechnic State University, 2008. http://ethics.calpoly.edu/ONR_report.pdf (accessed March 12, 2011.

May, Larry, Eric Rovie, and Steve Viner, eds. *The Morality of War: Classical and Contemporary Readings.* Upper Saddle River, NJ: Pearson Education, 2006.

Maze, Rick. "Draft Registration Should End, Lawmaker Says." *Army Times*, February 14, 2011. http://www.armytimes.com/news/2011/02/military-ending-draft-registration-021411w/ (accessed February 20, 2011).

McDaniel, Erin A. "Robot Wars: Legal and Ethical Dilemmas of Using Unmanned Robotics Systems in 21st Century Warfare and Beyond." Fort Leavenworth, KS: U.S. Army Command and General Staff College, 2008.

McGrath, Shaun R. *Strategic Misstep: "Immortal" Robotic Warfare, Inviting Combat to Suburban America.* Carlisle, PA: U.S. Army War College, 2010.

Mental Health Advisory Team IV. *Final Report.* Office of the Surgeon, Multinational Force–Iraq, and Office of the Surgeon General, U.S. Army Medical Command, 2007. http://www.armymedicine.army.mil/reports/mhat/mhat_iv/mhat-iv.cfm.

Moss, Kenneth B. *Undeclared War and the Future of U.S. Foreign Policy.* Washington: Woodrow Wilson International Center for Scholars and Johns Hopkins University Press, 2008.

Müller, Harald. "The Antinomy of Democratic Peace." *International Politics* 41, no. 4 (December 2004): 494.

National Research Council Board on Army Science and Technology. *Star 21: Strategic Technologies for the Army of the Twenty-First Century.* Washington: National Academy Press, 1992.

National Security Strategy: May 2010. Washington: Office of the President of the United States, 2010.

O'Connell, Mary Ellen. "Flying Blind." *America* 202, no. 8 (March 15, 2010): 10.

———. "Respect the Battlefield." CBS News, April 9, 2010. http://www.cbsnews.com/stories/2010/04/08/opinion/main6377556.shtml (accessed October 20, 2010).

———. "Unlawful Killing with Combat Drones: A Case Study of Pakistan, 2004–2009, Notre Dame Legal Studies Paper No. 09-43, Social Science Research Network, July 2010. http://papers.ssrn.com/sol3/papers.cfm?abstract_id=1501144 (accessed January 23, 2013).

Orend, Brian. "War." In *The Stanford Encyclopedia of Philosophy*, edited by Edward N. Zalta. Online edition, 2008. http://plato.stanford.edu/archives/fall2008/entries/war/ (accessed November, 2010)

Partridge, Kenneth, ed. *Robotics.* New York: H. W. Wilson, 2010.

Powell, Colin L., and Joseph E. Persico. *My American Journey.* New York: Random House, 1995.

Public Broadcasting Service. "The Invasion of Iraq: Operation Iraqi Freedom." *Frontline*, February 26, 2004, http://www.pbs.org/wgbh/pages/frontline/shows/invasion/cron/ (accessed January 30, 2011).

———. "Napoleon: Napoleon at War." 2000. http://www.pbs.org/empires/napoleon/n_war/campaign/page_12.html (accessed January 30, 2011).

Pylyshyn, Zenon W., ed. *The Robot's Dilemma: The Frame Problem in Artificial Intelligence.* Norwood, NJ: Ablex, 1987.

Record, Jeffrey. "Force-Protection Fetishism: Sources, Consequences, and (?) Solutions." *Air & Space Power Journal* 14, no. 2 (January 2000): 10.

"Remarks by Secretary Gates at the United States Air Force Academy." *U.S. Department of Defense Public Affairs Release,* (March 4, 2011). http://www.defense.gov/transcripts/transcript.aspx?transcriptid=4779 (accessed March 10, 2011).

Roberts, Adam, and Richard Guelff, eds. *Documents on the Laws of War*. 3rd ed. New York: Oxford University Press, 2000.

Ruskin, John. *The Crown of Wild Olive: Four Lectures on Industry and War*. New York: J. Wiley & Son, 1874.

Savage, Charlie. "Case of Soldiers Accused in Afghan Civilian Killings May Be Worst of Two Wars." *New York Times*, October 4, 2010.

Schwarzkopf, H. Norman, and Peter Petre. *It Doesn't Take a Hero: General H. Norman Schwarzkopf, the Autobiography*. New York: Bantam, 1993.

Shanker, Thom. "Gates Warns Against Wars Like Iraq and Afghanistan." *New York Times*, February 26, 2011.

Sidgwick, Henry. *The Elements of Politics*. London: New York, Macmillan, 1891.

Singer, P. W. "The Ethics of Killer Applications: Why Is It So Hard to Talk about Morality When It Comes to New Military Technology?" *Journal of Military Ethics* 9, no. 4 (December 2010): 299–312.

———. "Robots and the Rise of 'Tactical Generals.'" *Defense News* 24, no. 10 (March 2009): 29–29.

———. "War of the Machines." *Scientific American* 303, no. 1 (July 2010): 56–63.

———. *Wired for War: The Robotics Revolution and Conflict in the Twenty-First Century*. New York: Penguin, 2009.

Smith, Rupert. *The Utility of Force: The Art of War in the Modern World.* New York: Vintage, 2008.

Strozzi-Heckler, Richard. *In Search of the Warrior Spirit*. Berkeley, CA: North Atlantic Books, 1990.

Szasz, Ferenc Morton. *The Day the Sun Rose Twice: The Story of the Trinity Site Nuclear Explosion, July 16, 1945*. Albuquerque: University of New Mexico Press, 1984.

"Targeting Terror." *Current Events* 98, no. 1 (September 1998): 3.

Taylor, William W., J. H. Bigelow, and John A. Ausink. *Fighter Drawdown Dynamics: Effects on Aircrew Inventories*. Santa Monica, CA: Rand Corp., 2009.

Tesla, Nikola. *My Inventions: The Autobiography of Nikola Tesla*. Radford, VA: Wilder Publications, 2007.

Ticehurst, Rupert. "The Martens Clause and the Laws of Armed Conflict." *International Review of the Red Cross* 37, no. 317 (April 1997).

Time. "Iraqi Miscalculation," January 11, 1993, 10.

Trainor, Bernard E. "Amphibious Operations in the Gulf War." *Marine Corps Gazette* 78, no. 8 (August 1994): 56.

Trimble, Stephen. "Boeing Considers Restarting OV-10 Production after 23-Year Hiatus." Flightglobal, February 1, 2009. http://www.flightglobal.com/articles/2009/02/01/321730/boeing-considers-restarting-ov-10-production-after-23-year.html (accessed March 4, 2011).

van Creveld, Martin. *Technology and War: From 2000 B.C. to the Present*. New York: Free Press, 1991.

———. *The Transformation of War*. New York: Collier Macmillan, 1991.

von Clausewitz, Carl. *On War*. Michael Howard, Peter Paret, and Bernard Brodie, eds. Princeton, NJ: Princeton University Press, 1989.

Walzer, Michael. *Arguing about War*. New Haven, CT: Yale University Press, 2004.

———. *Just and Unjust Wars: A Moral Argument with Historical Illustrations*. New York: Basic Books, 2006.

Wrage, Stephen. "When War Isn't Hell: A Cautionary Tale." *Current History* 102, no. 660 (January 2003).

Zenko, Micah. *Between Threats and War: U.S. Discrete Military Operations in the Post–Cold War World*. Stanford, CA: Stanford University Press, 2010.

INDEX

Abney, Keith, 126, 143
accountability, 106, 136–46
  manufacturers, 139–41, 145–46
  military, 141–44, 145–46
  moral agency and rule of law, 127–28, 133
  overview, 136–39, 145–46
  programmers/designers, 139–41, 144–46
Adams, Thomas, 105
advantages of using lethal robotics, 132, 133
Afghanistan
  defense budget, 8, 113
  Operation Anaconda, 94
  Operation Enduring Freedom, 10, 51, 59, 62, 115–17, 152, 160
  U.S. bombing of in 1998, 67–68
AI (artificial intelligence). *See* artificial intelligence
air combat, about, 90, 130–32, 142
aircrew management problem, 98–100
ALARMs (air-launched antiradiation missiles), 18, 20
Albright, Madeleine, 77
all-volunteer force of U.S. military, 71–72
Alperovitz, Gar, 48
Alston, Philip, 70
Amara, Roy, 168
Amara's Law, 168, 169, 171
ambiguities and AI, 131–32
antipersonnel land mines, 20–21, 28–29, 127
Arguing About War (Walzer), 153
Arkin, Ronald C., 125, 161
  military control, 106
  research, 132–33, 135, 138, 159–60
  robotic definitions, 14, 15, 16
arms race in robotic technology, 116, 121–22
Army Field Manual (FM) 3–24, 112
artificial intelligence (AI)
  accountability and decision-making, 144–47
  ambiguities, 131–32
  definitions, 14, 16–17, 20
  distinction capabilities, 126–28
  experiments, 132–36
  views of and current state, 125–26, 137
assignments, pilot, 109
Association for Unmanned Vehicles Systems International (AUVSI), 15, 27
asymmetrical warfare, 115–18, 121–22
atrophy of skills, 85, 100, 102–3, 104–6, 107
"Attack of the Killer Robots" (Stoner), 75
automation and defining robots, 12–17, 21
autonomous weapons, defined, 17–18
autonomy
  defining robots, 12–18, 20–21
  distinction, 105–6, 133–34
  need for moral debate, 23–24
  responsibility, 140, 142–44, 146
  supervised, 16–17, 20, 103
AUVSI (Association for Unmanned Vehicles Systems International), 15, 27

Bacevich, Andrew, 69
Base Realignment and Closure program, 99
battlefield ethics, 159–60
Bekey, George, 126, 143
Berrigan, Frida, 69
bin Laden, Osama, 67, 68
biological warfare, 29, 41, 152
blind pursuit of victory, 150–53, 164, 175
Boise Weekly, 69
bombing of Hiroshima and Nagasaki, 47, 48, 66
Boyd, John, 40
Bradley, Omar N., 167
budget, national defense, 8, 76, 99, 112, 113, 156
Builder, Carl, 5
Bush, George W., 109, 151

Camus, Albert, 55, 72, 83, 90
Casey, George W., Jr., 75, 110–11, 112, 121
casualties
  aversion to, 66–67, 78–79 (*See also* risk, aversion to)
  combatant *vs.* noncombatant, 43, 44–45
centralized execution, 102
Cessna Caravan, 109
Challenger, 139–40
chance of success in just war theory, 51, 53–54, 117, 175
chemical warfare, 26
Christmas Bombing of 2009, 73
civilian authority over military, 63, 65, 76–77, 162, 174
civilian-military relationship, 63, 65, 75–79, 80–81, 174
Clausewitz, Carl von, 52, 69, 120, 153, 176
Clinton, Bill, 67
CNN effect, 92, 98, 163
Coker, Christopher
  casualty aversion, 78–79
  detachment from war consequences, 73–74, 93, 176
  ethics striving and morality, 121, 153, 157, 161–62, 164, 167
  impunity, 114, 116, 119–20
  warrior ethos and culture, 58, 84, 87, 88, 89, 90
combat to home life, 96–98
combatants, 83–84
  consent of, 57–60, 174
  innocence of, 51, 56–58, 60
Coming Robot Army, The (Featherstone), 138
commander accountability, 141–44, 145–46
communications, importance of speed in, 40–42
conduct following war, 50
conduct in war
  distinction, 126
  impunity, 63, 77–78, 81, 174
  just war theory, 50–51, 153
  risk, 54, 56
consecutive miracles, 4, 161
consent of combatants, 57–60, 174
consequences, detachment from war, 92–95
constraints-based systems, 131–33
control of robotic systems, levels of, 15–17, 22–23, 101–2, 105, 134
Cook, Martin
  asymmetric warfare, 115
  civil-military relations, 77
  impunity, 62, 67, 68
  morality and ethical striving, 26–27, 50, 161
  on U.S. military, 71
Counter Rocket Artillery Mortars (C-RAM), 2–3, 12
counterinsurgency, 26, 103, 112
Counterinsurgency (FM 3-24), 112
CQ Researcher, 55
C-RAMs (Counter Rocket Artillery Mortars), 2–3, 12
culture, warrior, 6, 84, 87, 95, 122
Cummings, Missy, 17, 101, 169
cyber warfare, 97
"Cybernetics" (Turing), 128

DARPA (Defense Advanced Research Projects Agency), 8, 10, 75–76, 100
decentralized execution, 102
decision-making

accountability, 141–42
autonomy, 15, 16–17, 20, 103, 105
importance of speed, 39–41
morality, 33–34, 36
risk, 50, 66, 76–77, 79
Unmanned Systems Integrated Roadmap, 9, 12–13
defense budget, national, 8, 76, 99, 112, 156
defense-related unmanned systems. *See* robotic systems in warfare
deficiencies of lethal robots, 135
defining robots, 11–18, 21–22, 100
Dennett, Daniel, 130, 137
Department of Defense (DOD), 17, 135, 171
Joint Robotics Program, 13–14, 15, 22, 113
Unmanned Systems Integrated Roadmap, 9–13, 22–23, 38–39, 86, 100, 104, 112–13
Desert Storm. *See* Gulf War
designer accountability, 139–41, 144–46
deskilling of the military, 85, 100, 102–3, 104–6, 107
detachment from war consequences, 92–95
determinism, 138–39
dimensions of impunity, 62–81
civilian-military relationship, 63, 65, 75–79, 80–81, 174
overview, 62–64, 79–81
psychology of citizens, 63, 71–75, 80–81, 98
psychology of state's leaders, 63, 65–71, 80–81
psychology of virtual warriors, 92–95, 98
warrior ethos, 63–64, 84–92, 95, 107–8
direct-combat autonomous systems. *See* robotic systems in warfare
directional nature of laws of war, 28–30, 34, 152
discrimination, 29–30, 126–28, 132–36, 144–45
disproportionate force, 116, 117–18, 135
distance in warfare, 38, 39, 42–43, 47, 92–93
distinction between combatants and noncombatants, 29–30, 126–28, 132–36, 144–45
double edge sword of superior firepower, 115–18, 119–20
double effect doctrine, 51, 54–56, 60, 137, 153–55, 174
Duke of Sung, 149–50, 153
Dunlap, Charles, 97, 108, 176

Eisenhower School for National Security and Resource Strategy. *See* ICAF; Industrial College of the Armed Forces
EM diagram, 40
ends justify the means, 150–53, 175
energy maneuverability theory, 40
Enigma, 128
ethics, 176. *See also* morality
advisory councils, 171
ethical striving, 153, 159–62
lethal autonomy, 13, 16
need for debate, 22–24, 100
research, 132–33, 145
Ethics and War in the 21st Century (Coker), 73–74, 167
ethos, warrior, 63–64, 84–92, 95, 107–8, 121, 176
evolution of weaponry, 38–43
exceptionalism, 157–58
existential nature of war, 64, 84, 86–92, 120
experiments, robotic, 125, 132–36, 145
explosive-ordnance disposal robots, 8, 9, 113

fallibility, 146, 150, 159–62, 175–76
Featherstone, Steve, 69, 132, 138, 142
fighting to preserve the force, 111
frame problem, 129–32, 135, 144–45
free will, 138–39, 144, 145, 146–47
Friedman, Thomas, 103

Gartner, Inc., 169–70
Gartner Hype Cycle, 169–70, 171

Gates, Robert, 110, 112, 114, 121
General Atomics, 76
Geneva Protocol, 127
geographic coincidence, 39, 43–49
Glosson, Buster, 66
Glymour, Clark, 129
Gomez, Mike, 83
GPS (global positioning system), 21
Green, Leslie, 28
Greiner, Helen, 27
grieving of fighters on the sidelines, 18
Gross, Michael, 56, 116–18, 121
Grossman, Dave, 84, 92, 94
Gulf War, 18–20, 40
  impunity, 47, 67, 93
  as a just war, 162–63
  risk and casualties, 43, 66
  unmanned systems use, 8, 19, 23, 135

Hague Convention and regulations, 27, 28, 127
Hall, J. Storrs
  accountability, 137, 139
  AI research, 125, 128, 132, 134–35, 147
  morality, 31
Hallion, Richard, 19
Hamburg, bombing of, 43, 94
Harak, G. Simon, 93–94, 98
HARMs (high-speed antiradiation missiles), 18, 20
Harper's Magazine, 69
Hayes, Patrick, 129, 130
Hellfire missiles, 37, 56, 59, 109
Hentoff, Nat, 156
hierarchical responsibility, 153–55, 165
Highway of Death, 163
Hiroshima bombing, 47, 48, 66
home soils, bringing war to, 97–98
Horner, Chuck, 66
human decision-making. *See* decision-making
humanitarianism, 117–19, 126
Hussein, Saddam, 1–2, 18

ICAF (Industrial College of the Armed Forces), 10, 11, 134
ICBMs (intercontinental ballistic missiles), 41, 49
IEDs (improvised explosive devices), 115–16, 121–22, 168
Ignatieff, Michael, 1
  accountability and responsibility, 159, 163
  impunity, 58, 72, 118
  just war, 159, 163
  morality, 30, 35–36, 118, 176
images of war, 74, 92, 94, 163–64
immaculate war, 62, 74, 159, 174
immunity of noncombatants
  distinction, 126–28, 144
  responsibility, 153–55, 158, 165
  risk, 45, 51, 174
impunity
  dimensions of impunity (*See* dimensions of impunity)
  spectra of, 42–49, 58, 60
In Search of the Warrior Spirit (Strozzi-Heckler), 87–89
Industrial College of the Armed Forces (ICAF), 10, 11, 134
industrial robots, 11–12, 14–15
industry of robotics technology, 10, 15–16, 100, 168, 171–72
infinite regression, 146–47
infrared (IR) air-to-ground missiles, 3–4
innocence of combatants, 51, 56–58, 60
instant news, 92, 98, 163
intentionality, 137–38, 144, 145, 146
intercontinental ballistic missiles (ICBMs), 41, 49
International Committee for Robot Arms Control, 23
international law, 25–26, 27, 34, 170
International Society for Military Ethics, 15–16, 17
Iraq War
  accountability, 141
  asymmetry, 115
  battlefield ethics, 159–60

command and control and decision-making, 19–20, 53, 102
risk aversion, 66–67
use of unmanned technology, 11, 42
iRobot, 76, 113
Israel, 44, 158

Japan, 61, 66, 78, 89
Johnson, Gordon, 25, 27
Joint Robotics Program, 13–14, 15, 22, 113
Joint Robotics Program Master Plan FY2005 (DOD), 13–15, 22
Journal of Military Ethics, 64
jus ad bellum. *See* resorting to war
jus in bello. *See* conduct in war
jus post bellum, 50
Just and Unjust Wars (Walzer), 33–34, 55
just war theory
accountability and responsibility, 127–28, 137, 153, 157, 163–64
impunity, 63, 67, 117
morality of risk and war, 26, 55–56, 59–60
unmanned warfare subverting criterion of, 168, 174–75
justification of war, 26, 63, 117

Kant, Immanuel, 1, 32, 133, 166
Kaplan, Robert, 69
Kierkegaard, Søren, 129
Killer Robots (Krishnan), 5, 14, 135
killing
distance/remote, 84–86, 87, 92–93, 96–97
with impunity, 57–58, 63, 69–70, 92, 98
of innocents, 54–55, 154, 174
resistance to, 31–34, 35, 92, 96, 162, 176
targeted, 44, 109, 116, 158
Koplow, David, 68, 79, 117, 118
Kosovo, war in, 159
impunity, 67–68, 96, 118
risk, 8, 43–44, 47–48, 52–53, 55
Krepinevich, Andrew, 86
Krishnan, Armin
autonomy, 13, 14, 15, 16
deficiencies of autonomous robots, 5, 135, 136
international law, 34
military skills and training, 105, 106
moral equality, 120
Kubrick, Stanley, 137
Kuwait, 18, 40, 71, 162

land mines, 12, 20–21, 28–29, 127
laws of war
accountability and responsibility, 127–28, 144, 146
legal combatants, 85, 168
limitations, 27–30, 34–36
morality, 25–27
remote killing, 70–71, 97–98
legality of unmanned systems, 12, 22–24, 27–28, 34–36, 57–58
Leonard, George, 87–88
lethal use of autonomous systems. *See also* robotic systems in warfare
accountability, 140, 142, 145–46
distinction and morality, 127–28, 145–46, 161, 168
human decision-making, 9, 11–13, 16, 41
research and need for debate, 13, 132–36
levels of control of robotic systems, 15–17, 22–23, 101–2, 105, 134
limit of war, practical, 163–64
Lin, Patrick, 71, 126, 143
Littoral Combat Ship, 11
living in two worlds, 96–98
Love, Maryann Cusimano, 95, 96
Lusso, Emilio, 33–34

M-160 robotic vehicles, 115–16, 135
Manhattan Project, 78, 100, 125–26
manning requirements, 9, 98–100, 104, 112–13
man-on-the-loop concept, 9, 17, 101, 105
manufacturer accountability, 139–41, 145–46

Mao Zedong, 149
Marshall, S. L. A., 31
Martens Clause, 27
McCarthy, John, 129
McDermott, Drew, 138–39
Mental Health Advisory Team for Operation Iraqi Freedom, 159–60
military
  accountability, 141–44, 145–46
  all-volunteer force of, 71–72
  control, 85, 101, 106, 107
  future of U.S., 109–12
Milne, A.A., 62
Milošević, Slobodan, 52–53, 55, 68
mines, 12, 20–21, 28–29, 127
Minsky, Marvin, 14
Mogadishu, 68, 119
moral agency, 15, 30, 127, 133, 145, 158, 165
moral equality, 133, 157, 168, 174
  impunity, 95, 118–21, 122–23
  risk, 56–58, 60
moral urgency, 151, 154, 164, 175
morality. *See also* ethics
  legality and, 30–34, 36, 167–68
  risk and responsibility, 51–58, 143, 173
  of unmanned systems, 8, 12
  of war, 25–28
Moshkina, Lilia, 14, 15
Moss, Ken, 65, 67
Müller, Harald, 77–78, 92

Nagasaki bombing, 48, 66
National Security Strategy of 2010, 156, 157
NDAA (National Defense Authorization Act) for Fiscal Year 2001, 7–9, 11, 38
New York Times, 44, 160
Nietzsche, Friedrich, 167
noncombatant immunity. *See* immunity of noncombatants
Normandy, 172–73
nuclear war, 26, 39, 45–46, 48–49
Nuremberg Trials, 30, 157

objective madman, 129
O'Connell, Mary Ellen, 69–70, 96
Office of Naval Research (ONR), 75, 133–35
On Killing: The Psychological Cost of Learning to Kill in War and Society (Grossman), 84
OODA Loop, 40–41
Operation Allied Force. *See* Kosovo, war in
Operation Anaconda, 94
Operation Desert Storm. *See* Gulf War
Operation Enduring Freedom, 10, 51, 59, 62, 115–17, 152, 160
Operation Iraqi Freedom. *See* Iraq War
Operations Northern and Southern Watch, 1–2, 102
Oppenheimer, J. Robert, 126, 147, 165
Orwell, George, 103, 138, 139
Ottawa Accord of 1997, 29, 34, 127
outward responsibility, 153–55, 158, 165
OV-10 Bronco, 110

PackBot, 76, 113
Pakistan, 29, 44, 70, 117, 119
Parameters (Adams), 105
passionate decisions, 158–61
past as war of the future, 110–15
Patriot missiles, 12, 18–20, 141
Patton, George S., 75, 120
persistence of battle, 39, 42, 110–15, 118, 120, 121–22
personnel requirements, 9, 98–100, 104, 112–13
Phalanx close-in ship-defense system, 3
politics of war, 63, 64, 65–71, 79, 174
post-traumatic stress disorder (PTSD), 95
Powell, Colin, 77, 163
precision in warfare, 42–43, 47, 52, 56, 67–68, 77, 79
Predators, 8, 9, 37, 41, 76, 87, 135, 168
preserve the force, fighting to, 111
Program Budget Decision 720, 99
programmer accountability, 139–41, 144–46
proportionality, 49, 51, 53–54, 60, 116–18, 135, 154

psychology
  of citizens, 63, 71–75, 80–81, 98
  of state's leaders, 63, 65–71, 80–81
  of virtual warriors, 92–95, 98
public opinion, 66–67, 73, 162–64, 174

radars as characteristic of electronic warfare, 19–20
radical responsibilities, 51, 150–58, 165
reactivity of law, 29, 35
Reapers, 8, 9, 41, 168
Record, Jeffrey, 52, 53
relevancy, 128–32, 136
remote warriors, 85, 92–95, 175
repeatability, 16
research, 75, 100, 125–26, 130, 132–36, 145, 171
resiliency of law, 29–30
resistance to killing, 31–34, 35, 92, 96, 162, 176
resorting to war
  impunity, 63, 67, 77–78, 81, 174
  risk, 50–51, 53–54, 56
responsibility. *See also* accountability
  autonomy, 140, 142–44, 146
  immunity of noncombatants, 153–55, 158, 165
  just war theory, 127–28, 153, 157
  radical, 51, 150–58, 165
restraint in warfare, 31–34, 118
right to be judged as a combatant, 57–58
risk
  aversion to, 48–49, 52–58, 59–60
  inherent in combatants, 50–51, 59, 90, 173
  inversion, 48–49, 59–60, 97, 116, 122, 154–55, 173
  morality of, 51–58
  spectra of impunity, 42–49
RMA (Revolution in Military Affairs), 77–78, 92
robot conscience, 133
robotic systems in warfare. *See also* lethal use of autonomous systems
  current state, 9–11, 168–70
  ethics of (*See* ethics)
  future of, 22–23, 168–72
  history, 7–9
  repair and maintenance, 113–14
Robotic Systems Joint Program Office (RS JPO), 113, 115–16
Robotics and Autonomous Systems Industry Study, 134
Robotics Technology Consortium, 10
robots, defined, 11–18, 21–22, 100
Robots in War: Issues of Risk and Ethics (Lin, Beckey, Abner), 126
ROE (rules of engagement), 106, 131–32, 160, 176
Rommel, Erwin, 56, 120
RPA operators as targets, 97–98
rules of engagement (ROE), 106, 131–32, 160, 176
Ruskin, John, 61, 79, 81

scenario fulfillment, 132
Schwarzkopf, H. Norman, 163
self-deterrence, 68, 118
sense-think-act paradigm, 14, 15, 16, 19, 21
Sharkey, Noel, 23, 125
Sherman, William Tecumseh, 176
Sidgwick, Henry, 25, 73, 75
Singer, Peter, 72, 102
  morality and moral equality, 64, 119
  robotics, 6, 7, 8, 14, 16, 168
  warrior ethos and culture, 86, 87
situational awareness, 10, 131–32
slave morality, 143
Smith, Rupert, 111, 114, 121
Socrates, 167
software, 3, 4–5, 19, 135, 161
Some Philosophical Problems from the Standpoint of Artificial Intelligence (McCarthy and Hayes), 129
Sparrow, Robert, 139, 143
Special Rapporteur on Targeted Killings and Extrajudicial and Summary Executions, 44, 70
spectra of impunity, 42–49, 58, 60
speed of warfare, 39–41, 42–43

standard language, need for in robotics, 22
Stimson, Henry, 61, 62
Stoner, Eric, 69, 75
Strozzi-Heckler, Richard, 87–89
submarine warfare, 29–30
success chances in just war theory, 51, 53–54, 117, 175
Sudan, 67–68
superior firepower as a double edged sword, 115–18, 119–20
supervised autonomy, 16–17, 20, 101, 103, 134

TAMI 21, 98–99, 106
targets, unmanned vehicle operators as, 96–98
technological trinity, 68, 79, 81
Technology and War: From 2000 B.C. to the Present (van Creveld), 39
technology competitions, 10
technology trends in warfare, 39–43
tension between winning and fighting well, 150–53, 164
Tesla, Nikola, 37
tests and evaluations of warfare technology, 3–5, 9, 10, 134–35
TLAMs (Tomahawk land attack missiles), 18–19, 21, 41, 67
training
  autonomy and accountability, 143, 144
  commander and officer, 99–101, 103–6, 107
  robot repair, 113–14
Transformational Aircrew Management Initiative for the 21st Century (TAMI 21), 98–99, 106
Tree, Sanho, 48
Trentonian, The (Hentoff), 156
Truman, Harry S., 78, 167
Turing, Alan, 128, 135
Turing Test, 128, 146
2001: A Space Odyssey, 137

UASs (unmanned aircraft systems), 8, 9–10, 13, 105, 169
UAVs (unmanned aerial vehicles), 17, 134, 135, 169
UGVs (unmanned ground vehicles), 8, 10, 13–14, 15, 134
UMVs (unmanned maritime vehicles), 10
Undeclared War and the Future of U.S. Foreign Policy (Moss), 65
Unmanned Aircraft Systems Flight Plan (USAF), 9, 13, 104, 105–6
unmanned systems in warfare. *See* robotic systems in warfare; *specific types*
Unmanned Systems Integrated Roadmap (DOD), 9–13, 22–23, 38–39, 86, 100, 104, 112–13
Urban Challenge, 10
USAF (United States Air Force), 5, 18, 86, 89, 102, 105
USVs (unmanned surface vehicles), 10, 101, 134
Utility of Force: The Art of War in the Modern World, The (Smith), 111
UUVs (unmanned undersea vehicles), 10–11, 134

van Creveld, Martin, 39, 175–76
venture capitalism, 75–76
Vietnam War, 8, 43, 66, 77
views of robots, utopian and apocalyptic, 100
virtual war, 72, 74, 92–95

Walzer, Michael, 1
  impunity and psychology of leaders, 65–66
  just war theory, 26, 90, 127, 163
  moral equality and moral urgency, 57, 151
  resistance to kill, 33–34, 96
  responsibility, 51, 153, 154–55, 156–57
  risk, 54–55
war, existential nature of, 64, 84, 86–92, 120
war-among-the-people trend, 111–12, 120, 164
"war in a can" concept, 132–33, 144

Warner, John, 7, 38
warrior, term of, 83–84
warrior culture, 6, 84, 87, 95, 122
warrior ethos, 63–64, 84–92, 95, 107–8, 121, 176
"When HAL Kills, Who's to Blame?" (Dennett), 137–38
When War Isn't Hell: A Cautionary Tale (Wrage), 77
Wired for War (Singer), 6, 7
World War II, 61, 66, 78, 89, 94, 156–57, 172–73
Wrage, Stephen, 77

Xenophon, 92

Yemen, 10, 37, 69–70, 114
Yugoslavia, 44

ABOUT THE AUTHOR

M. Shane Riza is a U.S. Air Force officer commissioned in 1990 upon graduation from the United States Air Force Academy with a degree in aeronautical engineering. A command pilot with 3,000 total flying hours, 2,800 of them in the F-16 multirole fighter, Colonel Riza is an experienced commander of both a squadron and a group. He is a graduate and former instructor of the USAF Weapons School, the Air Force's premiere tactical aviation school. In more than twenty years of service, he has been assigned in Asia, Europe, and the United States; has deployed to named operations on five occasions; and commanded a fighter squadron engaged in Operation Iraqi Freedom. He is a Distinguished Graduate of the Industrial College of the Armed Forces—now the Dwight D. Eisenhower School for National Security and Resource Strategy—at Fort McNair, in Washington, D.C., with a master of science degree in national security resource strategy. He holds two other master's degrees in military art and science. Colonel Riza has been a panelist at a National Defense University conference on unmanned warfare and has presented at the U.S. Army Command and General Staff College ethics symposium. Previous publications include "The Operational and Tactical Nexus: Small Steps Toward Seamless Effects-Based Operations," published as a Wright Flyer paper by Air University Press, and "A Grand Unified Theory of Fighter Quantum Mechanics: The Case for Air-to-Air Training in Multi-Role Fighters," which appeared in the Weapons School's official magazine, *The Weapons Review*.

M. Shane Riza continues to research and write on the subjects of morality, war, and the warrior ethos while dividing his time between his current assignment, his home state of Texas, and the north Georgia mountains.

For more information, please visit www.shaneriza.com.